# 5th International Conference on Civil Engineering and Materials Science (ICCEMS 2020)



Held online due to COVID-19

IOP Conference Series: Materials Science and Engineering
Volume 897

Singapore
15 - 18 May 2020

ISBN: 978-1-7138-2811-2
ISSN: 1757-8981

**Printed from e-media with permission by:**

Curran Associates, Inc.
57 Morehouse Lane
Red Hook, NY 12571

**Some format issues inherent in the e-media version may also appear in this print version.**



Institute of Physics
Dirac House, Temple Back
Bristol BS1 6BE UK

Phone: 44 1 17 929 7481
Fax: 44 1 17 920 0979

techtracking@iop.org

**Additional copies of this publication are available from:**

Curran Associates, Inc.
57 Morehouse Lane
Red Hook, NY 12571 USA
Phone: 845-758-0400
Fax: 845-758-2633
Email: curran@proceedings.com
Web: www.proceedings.com

# TABLE OF CONTENTS

## MATERIALS SCIENCE AND ENGINEERING



## DESIGN AND ANALYSIS OF BUILDING MATERIALS

## APPLIED MECHANICS



## CIVIL AND ARCHITECTURAL ENGINEERING

IOP Conf. Series: Materials Science and Engineering **897** (2020) 011001 doi:10.1088/1757-899X/897/1/011001

# Preface

The 5th International Conference on Civil Engineering and Materials Science (ICCEMS 2020) was held in May 2020. Due to the outbreak of COVID-19, this year's conference, which was supposed to be held in Singapore, was organized as fully virtual conference, in meeting system "Zoom". In order to prepare ICCEMS 2020, the organizing committees, reviewers, session chairs as well as all the authors and presenters have made a lot of efforts and contributions. Thank you very much for always being supportive to the conference. We could not have pulled off this convention without all of your hard work and dedication.

Since their inception, ICCEMS have been attracting many delegates from all over the world. The conference was consisted of 3 keynote speeches, from Japan, Australia and China, 4 technical sessions, over 30 representatives from more than 8 countries, for example, China, Philippines, Germany, Malaysia, Thailand, Indonesia, USA, India and etc.

Each keynote speech lasted for 45min including 5min for Q&A, while for the authors' presentation, 15min including Q&A was allocated apiece. All participant were active, questions raised after every talk. The whole conference was held online in ZOOM, we also had test for each presenters in advance to ensure the success of delivery of the conference. For those who had internet problem, we also accepted a pre-recorded video presentation as an alternative of online presentation. Meanwhile the whole conference had been recorded for conference back-up only.

ICCEMS aims to become premier international conference for in-depth discussion on the most-up-to-date and innovative ideas, research projects and practices in the field of civil engineering and material sciences. Papers published in the ICCEMS 2020 proceedings cover various topics from civil and structural engineering to nanomaterials science as well as a broad range of related interdisciplinary subjects.

Today, the world develops with unprecedented speed, which is only possible with advances in various fields of sciences and engineering. We are very pleased to contribute to this important global effort. We sincerely hope that the academic community and industrial practitioners will continue to support us in our attempts to provide even more meaningful conferences with numerous critical idea exchanges, diverse opportunities for fruitful networking and future collaborations between the delegates.

Warmest regards,

Conference Chair
Prof. Xiaohong Zhu
Sichuan University, Chengdu, China
ICCEMS 2020 Committee

IOP Conf. Series: Materials Science and Engineering **897** (2020) 011002 doi:10.1088/1757-899X/897/1/011002

# Conference Committees

**International Advisory Committees**

Richard G. Haverkamp, Massey University, New Zealand

Tom Wu, The University of New South Wales, Australia

**Conference Chair**

Xiaohong Zhu, Sichuan University, China

**Steering Committees**

Kyoung Sun Moon, Yale University School of Architecture, USA

C. W. Lim, City University of Hong Kong, Hong Kong

**Program Chair**

Mohd Rafie Johan, University of Malaya, Malaysia

**Publicity Chair**

Chien-Yuan Chen, National Chiayi University, Taiwan

**Technical Program Co-Chair**

Shougen Chen, Southwest Jiao Tong University, China

**Technical Committees**

Xiaojun Li, California State University, Fresno, USA

Osama Ahmed Mohamed, Abu Dhabi University, UAE

Hanif Heidari, Damghan University, Iran

Servando Chinchón Yepes, University of Alicante, Spain

Shantha Kumari Muniyandi, Tunku Abdul Rahman University College, Malaysia

Wan Inn Goh, Universiti Tun Hussein Onn Malaysia, Malaysia

Kong Fah Tee, University of Greenwich, UK

Muttaqin Hasan, Syiah Kuala University, Indonesia

Haryati Yaacob, Universiti Teknologi Malaysia, Malaysia

Junting Jiao, Xiamen University of Technology, China

# Peer review declaration

All papers published in this volume of *IOP Conference Series: Materials Science and Engineering* have been peer reviewed through processes administered by the Editors. Reviews were conducted by expert referees to the professional and scientific standards expected of a proceedings journal published by IOP Publishing.

- **Type of peer review:** Double-blind
- **Conference submission management system:** Easychair
- **Number of submissions received:** 33 (30 full papers+ 3 abstracts)
- **Number of submissions sent for review:** 30
- **Number of submissions accepted:** 17
- **Acceptance Rate (Number of Submissions Accepted / Number of Submissions Received X 100):** 51.5%
- **Average number of reviews per paper:** 3
- **Total number of reviewers involved:** 51
- **Any additional info on review process:**

  After received, all submissions were subjected to double-blind peer reviews, who are expert or have been experiencing in the related field for years. The accepted papers must be revised, taking into consideration the referees' comments and suggestions, before inclusion in the conference proceedings. Manuscripts should be original (not submitted/published anywhere else) and written in accordance with the standard IOP paper template.

- **Contact person for queries:** Mr. Quanming Yuan(qm.yuan@zhconf.ac.cn)

# Synthesis of Irregular Tantalum Pentoxide ($Ta_2O_5$) Microparticles by Direct Thermal Oxidation of Ta Foils in Atmospheric Oxygen

**D. Pérez-Pérez [a], R. Acosta-Vera [a], A. O. Ramírez-Morales [a], Oscar E. Cigarroa-Mayorga [a,*]**

[a]Dept. Advanced Technologies, UPIITA-Instituto Politécnico Nacional, Av. IPN 2580, Col. Ticomán, Gustavo A. Madero, C. P. 07360, CDMX, México.

*Corresponding author's e-mail: ocigarroam@ipn.mx (Oscar E. Cigarroa-Mayorga)

**Abstract.** In this work, Tantalum pentoxide ($Ta_2O_5$) microparticles were achieved by one-step thermal oxidation of Tantalum (Ta) foils under 600 °C for 6 h. The effect of time and temperature was explored in the process (50 to 600 °C, 1 to 6 h). Thus, Ta foil was chemically cleaned with nitric and hydrochloric acid and then was immersed in water; finally it was dried with Nitrogen flux. The foils were isothermally oxidized in a horizontal furnace with 21 wt. % Oxygen atmosphere in a constant flux of 20 L/min. The x-ray diffraction, scanning electron microscopy, Raman and energy dispersive x-ray spectroscopy led to the conclusion that formation mechanism of microparticles is mainly constituted by lattice parameters mismatching between Ta and $Ta_2O_5$ and crystallite size difference. Along the oxidation process at 600 °C, the phase transformation across time carries defect concentrations on entire material experimentally demonstrated by $Ta_2O_5$ Raman shift and then the sample is broken in irregular shape. As oxidation time reaches 6 h value, phase transformation in Ta foil causes mismatching that firstly diminishes crystallite size, then evolves to cracks and finish by the formation of the particles in the micrometer order size.

**Keywords:** phase transformation, $Ta_2O_5$ Raman shift, formation mechanism, lattice mismatching

## 1. Introduction

Tantalum pentoxide ($Ta_2O_5$) is a biocompatible and semiconductor material. It has been proved to have many promising applications such as corrosion resistance covering material for biomedical applications [1] and dielectric applications by a thin film on semiconductors [2]. Among different morphologies, Tantalum oxide particles have potentials applications for angiography as well as RES-targeted imaging to improve the detection of metastases in the liver and lymph nodes [3]. It is possible to make compounds with $Ta_2O_5$ to improve the surface bio-performances of polyimide (PI), the surface roughness, hydrophilicity, surface energy and protein adsorption as well as apatite mineralization of STPC in simulated body fluid were significantly enhanced [4]. Gram-scale synthesis of uniform-sized TaOx particles was achieved using a simple microemulsion method in the work of Myoung Hwan et al [4]. As it was reported by Krishnaprasanth and Seetha $Ta_2O_5$ particles can be prepared by thermal decomposition method without using any solvents [5]. $Ta_2O_5$ nanoparticles with the size in the range of 20-50 nm were obtained via a chemical route described in the work of Huang and Hsieh [6]. Among methodology to synthesized metal-base micro-structured materials, heterogeneous nucleation promoted by thermal oxidation has been proved to be an accessible method to produce uniform oxides-phase from highly pure metal foils in an extensive area. [7]

This research contributes to the exploration of accessible methodologies for obtaining highly pure microparticles with a single orthorhombic tantalum pentoxide phase. In this work, a Tantalum (Ta) foil was chemically cleaned with hydrochloric and nitric acid solutions, then Ta foil was rinsed in water and dried with Nitrogen flux. Afterword, the plate was thermally oxidized in a 21 wt. % Oxygen atmosphere for 1 to 6 h at 50 °C to 600 °C range. $Ta_2O_5$. microparticles obtained at 600 °C were structural and morphologically characterized. The formation mechanism of $Ta_2O_5$.microparticles is discussed based on results.

## 2. Materials and methods

### *2.1 Chemicals*

The Tantalum (Ta, 99.99 %), nitric acid ($HNO_3$, 70 %), hydrochloric acid (HCl, 70 %), deionized water ($H_2O$, 18 MΩ) and methanol were commercially acquired from Sigma Aldrich and used without further purification.

### *2.2 Microparticles synthesis*

To obtain Tantalum oxide, the Ta foil was cut small sheets (0.5 x 0.5 cm). Afterward, the Ta sheets were chemically cleaned with HCl 0.1 M solution for 1 min to remove organic impurities and then with $HNO_3$ 0.1 M solution for 1 min to remove metallic oxide impurities, finally deionized water was employed to remove residues of acid solution and then dried with Nitrogen flux to avoid further oxidation. Once Ta foils were cleaned, they were oxidized in a horizontal furnace in a 21 wt. % Oxygen atmosphere at constant rate of 20 L/min. In order to study the effect of heat in phase structure, a range of temperature from 50 to 600 °C was employed in the synthesis with 50 °C as increasing step. The effect of time in morphological properties was studied from 1 h to 6 h with a 30 min as increasing time. To study the reproducibility, each condition of synthesis was performed by triplicate in separated experiments.

### *2.3 Characterization*

The phase in samples was characterized by x-ray diffraction (XRD) using a x´pert Panalytical diffractometer (Cu kα = 1.5418 Å), while morphological properties of the samples were obtained by electron microscopy (SEM) using a JSM-7800F field emission scanning electron microscope (FE-SEM). To determine the pollutants present in micro-particles, an energy dispersive x-ray spectroscopy (EDS) was performed by an Eurotherm coupled system in the FE-SEM. The defects present in the condition of synthesis were analysed in the corresponding vibrational modes in Raman spectroscopy recorded by INTEGRA Spectra equipment with wavelength excitation of 532 nm.

## 3. Results

The structural features of samples after thermal oxidation on Ta foils were analysed by XRD pattern (Figure 1). Based on temperature exploration, it was observed that cubic Ta phase remains present till 300 °C is reached. The pattern obtained for samples processed between 50 °C and 300 °C is shown in Figure 1a. For this pattern, cubic Ta phase was identified by matching the picks at 38.47, 55.55 and 69.60 degrees with (011), (002) and (112) planes, respectively (Reference code: 98-007-6014). Single orthorhombic $Ta_2O_5$ phase was achieved when samples were processed at 600 °C. It was confirmed by matching pick position at 22.84, 28.29, 36.66, 46.66, 49.78, 50.68, 55.74, 58.52, 63.65 and 71.34 degrees with (100), (011), (111), (200), (020), (013), (202), (022), (122) and (213) planes for the corresponding phase in the pattern shown in Figure 1b, respectively (Reference code: 98-004-3498). The lattice parameters were computed by Rietveld approaching in High Score Plus software, obtaining as results for the Ta phase a set of values of a = b = c = 3.30 Å and for the $Ta_2O_5$ phase a collection of values of a = 3.89 Å, b= 3.66 Å and c= 6.20 Å. Based on these results, lattice for each face was simulated (inserts in Figure 1). As it can be seen, the full width at the half maximum (FWHM) in the

$Ta_2O_5$ phase is bigger than that from the Ta phase, thus not just the lattice parameters have great difference but crystallite size is smaller in $Ta_2O_5$ phase.

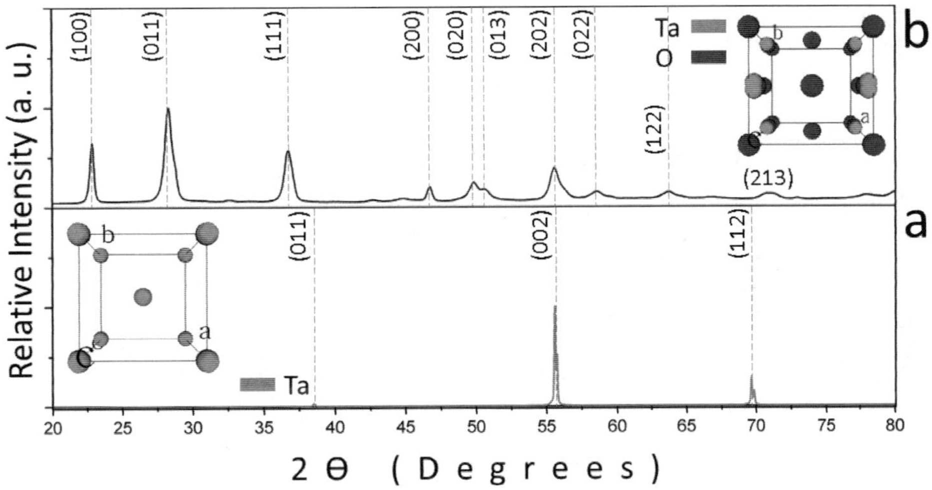


**Figure 1.** The normalized XRD pattern of samples thermally oxidized from (**a**) 50 to 300°C and (**b**) at 600 °C. Inserts show the simulated unit cell for (**a**) Tantalum and (**b**) orthorhombic tantalum pentoxide, respectively.

Samples with single $Ta_2O_5$ phase (obtained at 600 °C) were chosen to by characterized by FE-SEM (Figure 2). The morphological properties obtained after 1 h of temperature exposition, results in nano-pattern on surface with high crystallinity (Figure 2a and 2d), these is in good agreement with the reports [2]. Then, after 3 h of oxidation process, it can be seen microstructure changes on surface, crack appearance (pointed by blue arrow in Figure 2b) and pore generation (pointed yellow circles in Figure 2e). After 6 h of treatment the foil has completely cracked, allowing the formation of microparticles (Figure 2c). By analyzing the surface of one $Ta_2O_5$ microparticle, boundaries have a regular morphology (pointed by blue arrow in Figure 2f). This suggests that material cracking to drive particles formation is due to these particular sites, thus they play an important role in process.

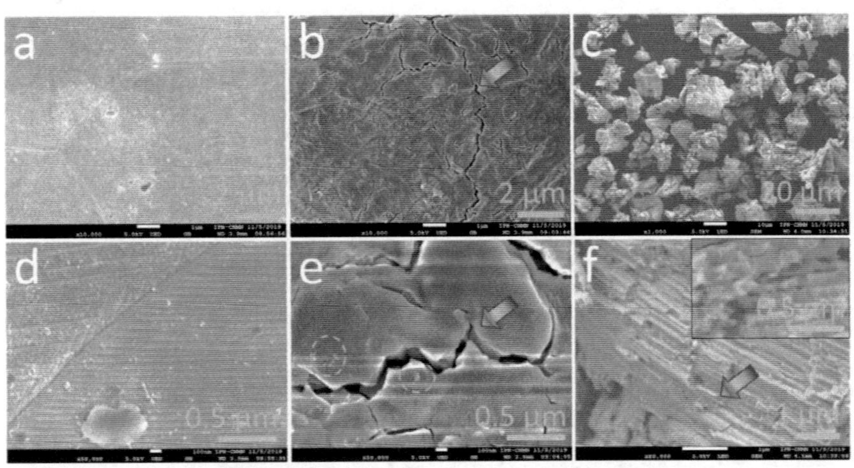


**Figure 2.** Ta foil imaged by FE-SEM after oxidation at 600 °C for (**a**) 1 h, (**b**) 3 h and (**c**) 6 h. The (**d**), (**e**) and (**f**) show close view of (**a**), (**b**) and (**c**) images, respectively. Insert in (**f**) is an amplified section.

EDS was performed on an extensive area of samples. As it can be seen in Figure 3a the energy values for Ta and O are the unique elements present in the samples (besides that from Carbon due to tape to hold the sample in FE-SEM). Raman spectroscopy is shown in Figure 3b, where the Raman shift are in good agreement with those reported for $Ta_2O_5$ in literature (see Table 1) [8]. The modes associated with the O – Ta – O and O – 3Ta binding are located at range from 150 to 400 $cm^{-1}$ (i and ii), to O – 3Ta stretching modes are located between 450 and 800 $cm^{-1}$ (iii, iv and v) and those associated to O – 2Ta stretching appear from 800 to 1000 $cm^{-1}$ (vi and vii). Vibrational modes shifting between samples in synthesized at different times, shows diverse defect amount in similar way it has been prove [9].

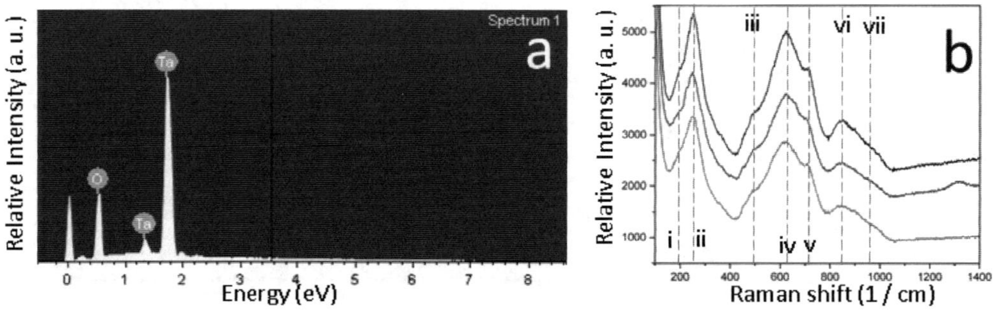


**Figure 3.** The (**a**) EDS spectra and (**b**) Raman shift of samples synthetized at 600 °C for 1 (red), 3 (blue) and 6 h (black).

**Table 1.** The Raman shift of Ta2O4 samples synthetized at 1, 3 and 6 h.

| Vibrational mode | 1 h ($cm^{-1}$) | 3 h ($cm^{-1}$) | 6 h ($cm^{-1}$) |
|---|---|---|---|
| i | 198.17 | 196.51 | 199.84 |
| ii | 251.42 | 246.43 | 248.76 |
| iii | 487.72 | 496.04 | 491.05 |
| iv | 612.52 | 619.18 | 620.84 |
| v | 707.38 | 702.38 | 712.37 |
| vi | 843.83 | 845.49 | 850.48 |
| vii | 950.33 | 958.65 | 973.62 |

The formation mechanism of microparticles can be explained based on results. First, when $Ta_2O_5$ nucleus are formed, defect induction and inner stress start in Ta crystal (Figure 4a), then $Ta_2O_5$ phase grows and drives defect concentration (Figure 4b), once phase transformation has finished polycrystalline $Ta_2O_4$ is resulting with inner stress due to defect concentration generated previously, thus material cracks and leave the formation of micro-particles.

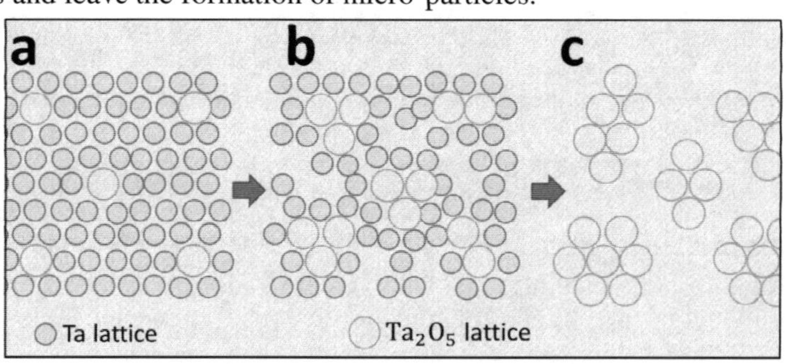


**Figure 4.** Scheme of the formation mechanism stages of $Ta_2O_5$ microparticles synthesized at 600 °C across time (**a**) initial, (**b**) middle and (**c**) final stage.

## 4. Conclusion

In summary, $Ta_2O_5$ microparticles were synthesized successfully by simply thermal oxidation on Ta foil. The $Ta_2O_5$ microparticles with a homogeneous orthorhombic phase were achieved at 600 °C and 6 h. According to vibrational spectroscopy results, the $Ta_2O_5$ formation mechanism is strictly related to defect concentration due to lattice mismatching between Ta and $Ta_2O_5$ in the oxidation process. This work gives a proposal for accessible methodology for obtaining $Ta_2O_5$ microparticles. It is recommended that future work stay focused on introduce an extra parameter to take control of size and shape.

## Acknowledgments

The authors acknowledge the *Secretaría de Investigación y Posgrado* of the Mexican *Instituto Politécnico Nacional (IPN)* for economic support to this work and the *centro de nanociencias y micro y nanotecnologías* of the *IPN* to the facilities for physical characterizations. The authors declare no conflict of interest.

## References

[1] S. Ying-Sui, C. Jean-Heng, and H. Her-Hsiung, "Corrosion resistance and biocompatibility of titanium surface coated with amorphous tantalum pentoxide" Thin Solid Films, Vol. 528, pp 130-135, January 2013.

[2] C. Chaneliere, J.L. Autran, R.A.B. Devine and B. Balland, "Tantalum pentoxide (Ta2O5) thin films for advanced dielectric applications", Materials Science and Engineering: R: Reports, Vol. 22, no 6, pp 69-322, May 1998.

[3] M. Hwan Oh, N. Lee, H. Kim, S. Pyo Park, Y. Piao, J. Lee, S. Woojoo Jun, W. Kyung Moon, S. Hong Choi and T. Hyeon, "Large-Scale Synthesis of Bioinert Tantalum Oxide Nanoparticles for X-ray Computed Tomography Imaging and Bimodal Image-Guided Sentinel Lymph Node Mapping", Journal of the American Chemical Society, Vol. 133, no 14, pp 5508–5515, March 2011.

[4] S. Asadullah, S. Mei, D. Wang, Y. Yao, Y. Pan, D. Wang, H. Guo, and J. Wei, "Sulfonated porous surface of tantalum pentoxide/polyimide composite with micro-submicro structures displaying antibacterial performances and stimulating cell responses" Materials & Design, Vol. 190, pp 108510, May 2020.

[5] A. Krishnaprasanth, and M. Seetha, "Solvent free synthesis of Ta2O5 nanoparticles and their photocatalytic properties". AIP advances, Vol. 8, no 5, pp 055017, April 2018

[6] H. C. Huang, and T. E. Hsieh, "Preparation and characterizations of tantalum pentoxide (Ta2O5) nanoparticles and UV curable Ta2O5 acrylic nanocomposites" Journal of Applied Polymer Science, Vol. 117, no 3, pp 1252-1259, March 2010.

[7] H. Rojas-Chávez, H. Cruz-Martínez, F. Montejo-Alvaro, R. Farías, Y.M. Hernández-Rodríguez, A. Guillen-Cervantes, A. Ávila-García, N. Cayetano-Castro, D.I. Medina, and O.E. Cigarroa-Mayorga, "The formation of ZnO structures using thermal oxidation: How a previous chemical etching favors either needle-like or cross-linked structures" Materials Science in Semiconductor Processing, Vol. 108, pp 104888, March 2020.

[8] C. Joseph, P. Bourson, and M.D. Fontana, "Amorphous to crystalline transformation in Ta2O5 studied by Raman spectroscopy" J. Raman Spectrosc., Vol. 43, no 8, pp 1146-1150, June 2020.

[9] Q. Li, H. Zhang, B. Cheng, R. Liu, B. Liu, J. Liu, Z. Chen, B. Zou, T. Cui, and B. Liu, "Pressure-induced amorphization in orthorhombic Ta2O5: An intrinsic character of crystal" Journal of Applied Physics, Vol. 115, no 19, pp 193512, May 2014.

IOP Conf. Series: Materials Science and Engineering **897** (2020) 012002 doi:10.1088/1757-899X/897/1/012002

# Ettringite dependace in thaumasite formation

**S Chinchón-Payá[1], A Aguado[2] and S Chinchón[3,4]**

[1] Instituto de Ciencias de la Construcción Eduardo Torroja (IETcc-CSIC), Madrid (Spain)
[2] Dept. of Construction Engineering, Universidad Politécnica de Cataluña, BarcelonaTech, (Barcelona, Spain)
[3] Dept. of Architectonic Constructions, Universidad de Alicante, (Alicante, Spain)
[4] Corresponding author's email: s.chinchon@ua.es

**Abstract.** Thaumasite has been synthetized with and without ettringite in the reaction medium. The compositions of solids have been analyzed over time by means of X-Ray Diffraction (XRD) and Scanning Electron Microscopy (SEM-EDX) in order to check the relation between ettringite and thaumasite. The results indicate that in the synthesis in absence of ettringite, the thaumasite appears at 6 months. By contrast, in the synthesis with presence of ettringite, at 7 days much of the ettringite has become thaumasite and at 2 months ettringite has disappeared and only thaumasite is detected. The results obtained at the end of two years experience, using the Rietveld method, show that thaumasite forms from ettringite by substitution of aluminum with silicon.

## 1. Introduction

Thaumasite is a mineral formed by calcium carbonates, silicates and sulfates $Ca_3Si(OH)_6(SO_4)(CO_3)\cdot 12H_2O$ and the interest in its study in cement based materials is relatively recent and topical [1-5]. Cases of roads and public highways where an abundant presence of thaumasite was detected in highly degraded concrete started to appear in the UK during the late 1990s [6], which led to the creation of a panel of experts from different institutions that were entrusted with the study of this issue [7]. The British interest materialized in the organization of a monographic conference in June 2002, [8] to which numerous case studies and research works were presented. The texts discussed during that conference shape an interesting themed volume with over 50 articles, which constitutes a very useful source of material when it comes to studying thaumasite [9].

Thaumasite is morphologically and structurally very similar to ettringite $3CaOAl_2O_33CaSO_4.32H_2O$ [10, 11] which explains why interpretation mistakes have often been made, identifying as ettringite what could actually be thaumasite. These mistakes may occur after rapidly interpreting X-ray diffraction spectra, since the characteristic peaks of both minerals are very close to one another.

Whereas ettringite formation has been extensively studied and ettringite is widely believed to form through a reaction of sulfates with the hydrated calcium aluminates of cement, a certain degree of controversy still exists around the formation of thaumasite and its possible connection with ettringite. Nevertheless, the current state of knowledge can roughly be grouped together around the three following theories [12, 13]:

1) According to the first theory, thaumasite is formed through ettringite by substitution of aluminum for silicon.
2) In the second case, ettringite acts as a nucleator for the thaumasite formation.
3) According to the third theory, thaumasite could be formed directly without influence of ettringite in the process at all.

This paper aims to present an experiment designed to test the influence of ettringite in the formation of thaumasite. The experience is based on a synthesis of thaumasite in the absence and presence of ettringite (hereafter Th series and Ett series respectively).

Several studies have been conducted synthesizing thaumasite in the laboratory. While Aguilera *et al.* [14] carried out a direct synthesis using pure reagents in stoichiometric proportions; the study by Kohler, Heinz [15] utilizes synthetic cement clinker phases, fly ash and nanosilica as pastes in an aqueous solution, to which calcite and gypsum or sodium sulfate are added. Of these two possibilities, the synthesis proposed by Aguilera, Varela [14] was used as a reference and basis of our work, as it is easier to control because it uses pure reagents and seems to get a higher concentration of thaumasite.

The studies carried out by Barnett *et al.* [16-18] succeeded in drawing a distinction between ettringite and thaumasite by cell parameter refinement of the structure in these two minerals. In fact, they prefer to consider compounds as ettringite-type and thaumasite-type depending on the structural similarities with pure crystals, in the belief that a gap or discontinuity exists in the solid solution, that ettringite allows Si to replace up to 1/2 of its Al atoms, and that thaumasite can tolerate little or no replacement of Si with Al atoms.

## 2. Materials and methods

The thaumasite synthesis procedure suggested by Aguilera *et al.* [14] (*Th series*) consists of the following steps: (1) preparing a *solution A* of CaO in an aqueous solution; (2) preparing a *solution B* with the rest of the necessary components/elements, that is, $Na_2SiO_3$, $Na_2CO_3$ and $Na_2SO_4$. A certain amount of sucrose is subsequently added to both solutions in order to avoid an excessive precipitation of by-products –unwanted secondary calcium compounds which would strongly reduce the performance of this reaction; and (3) these two solutions are mixed inside a polyethylene container hermetically closed before introducing it in the fridge at 5°C. The concentrations of the different reagents in the mix are as follows: 0.025 M CaO, 0.006 M $Na_2SiO_3$, 0.006 M $Na_2CO_3$ and 0.01 M $Na_2SO_4$. After some time, the samples are taken out of the fridge, filtered, dried and analyzed.

Instead, the modification through modified thaumasite synthesis (*Ett series*) consisted in adding previously synthesized ettringite to the medium through a process based on the method developed by Cody, Lee [19]. Two solutions, C and D, were prepared for this purpose: a *Solution C* with CaO in distilled water (0.185 gr CaO/100 mL $H_2O$) and a *Solution D* with $Al_2(SO_4)_3$ in distilled water (0.2317 gr $Al_2(SO_4)_3$/100 mL $H_2O$). Both solutions are mixed and the resulting precipitate is filtered.

Five reaction vessels were subsequently prepared following the same thaumasite synthesis process through the *Th series* (total 800 mL of solution in 1 L containers) but 1.00 gr of the ettringite synthesis sample was added on this occasion. The containers were hermetically closed and introduced in the fridge at 5°C.

Precipitates from both the *Th series* and *Ett series* were taken out of the fridge after timespans of 1 day, 7 days, 2 months, 6 months and 2 years, and later filtered and dried.

The following techniques and methods were applied to characterize the samples and analyze the results:

- SEM-EDX analyses were carried out with a scanning electron microscope Hitachi S3000N model. This microscope has an X-ray detector Bruker brand XFlash 3001 model for microanalysis and mapping.
- XRD of the experimental samples was conducted in a PANalytical X´Pert PRO dust diffractometer using $K\alpha 1$ radiation, a primary focusing monochromator, and an X´Celerator detector. $2\theta/\theta$ scans were carried out between 4 and 80 °2θ with a step size of 0.017 °2θ and a measuring time of 50 seconds per step.

In the case of the *Ett series* sample after 2 years, the Rietveld method [20, 21] served to refine the spectrum seeking to accurately determine the cell parameters of thaumasite. Rietveld refinements have been performed considering the 3 identified phases of sample (calcite, thaumasite and sucrose) by means of FullProf software [22].

## 3. Results and discussion

### *3.1. Thaumasite synthesis in absence of ettringite. Th series*

The crystalline phases identified by XRD in the samples of the *Th series* after 1 day, 7 days, 2 months, 6 months and 2 years are shown in Table 1. As we can see, only different calcium carbonates are detected during the first two months, namely: calcite $CaCO_3$; vaterite $CaCO_3$; and monohydrocalcite $CaCO_3.H_2O$; and thaumasite first appears –together with calcite– after 6 months (Fig. 1d). This is exactly in keeping with the observations of Aguilera *et al.* [14], according to which the *Th series* synthesis experience can be considered finished at that age.

**Table 1.** Identified compounds in the samples of *Th series*.

| | Calcite | Vaterite | Monohydrocalcite | Thaumasite |
|---|---|---|---|---|
| 1 day | ■ | ■ | | |
| 7 days | ■ | ■ | ■ | |
| 2 months | ■ | ■ | | |
| 6 months | ■ | | | ■ |
| 2 years | ■ | | | ■ |

In Figure 1 the SEM images of calcite, vaterite, the monohydrocalcite and thaumasite are shown. Figure 1 A, B, C show calcium carbonates polymorphs and figure 1D shows the first thaumasite crystals.

**Figure 1.** SEM images of calcite (A), vaterite (B), monohydrotalcite (C) and thaumasite on calcite crystals (D). (A, B, C) at 7 days and (D) at 6 months.

### *3.2. Thaumasite synthesis in presence of ettringite. Ett series*

Ettringite was first synthesized by using the method of Cody, Lee [19] described above. The resulting product was analyzed by XRD, and a concentration of 56% for ettringite and 44% for calcite was

IOP Conf. Series: Materials Science and Engineering **897** (2020) 012002 doi:10.1088/1757-899X/897/1/012002

obtained after applying the values of the I/Ic constant for ettringite (2.3) and calcite (3.2) published by International Center for Diffraction Data [23].

In Table 2 are represented crystalline phases identified by XRD of the Ett series samples obtained at timespans of 1 day, 7 days, 2 months, 1 year and 2 years. We can see that after 7 days ettringite has become thaumasite and at 2 months ettringite has disappeared transformed into more thaumasite crystals. The vaterite transforms into a more stable polymorph (calcite).

**Table 2.** Identified compounds in the samples of *Ett series*.

| | Calcite | Vaterite | Ettringite | Thaumasite |
|---|---|---|---|---|
| 1 day | ■ | ■ | ■ | |
| 7 days | ■ | ■ | ■ | ■ |
| 2 months | ■ | | | ■ |
| 6 months | ■ | | | ■ |
| 2 years | ■ | | | ■ |

*3.3. Ettringite-thaumaste relation*

Figure 2 provides the 5-17 °2θ zones of the three diffractograms for the *Ett series* which best explain the ettringite-thaumasite relation: 1 day, 7 days and 2 years.

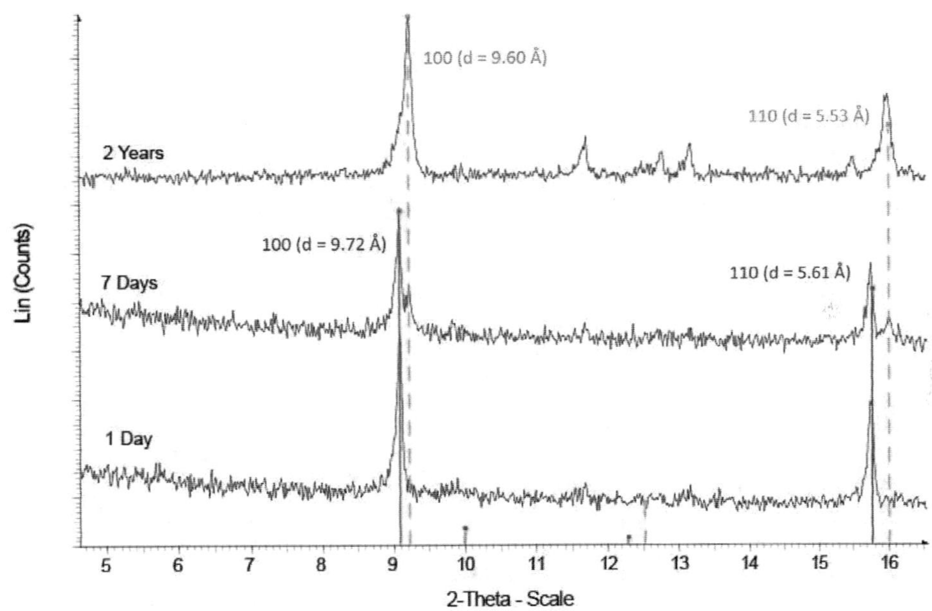


**Figure 2.** Diffractograms of *Ett series* at 1 day, 7 days and 2 years. The peaks belonging to ettringite and thaumasite are show in red and green respectively. Principal reflections and d-spacing values of principal peaks are also shown.

To check whether the thaumasite obtained at the end of the process of *Ett series* is a pure compound, or whether it is an end member of the thaumasite-ettringite solid solution, a refinement of the cell parameters of the sample at 2 years was made by using the Rietveld method. **Table 3** shows the cell parameters values of thaumasite obtained at 2 years in Ett series. The parameter "a" value turns out to be situated below 11.11 Å which could correspond to a thaumasite-ettringite solid solution compound [24].

**Table 3.** Refinement of cell parameters of the thaumasite detected in the mortar sample and thaumasite synthesized after 2 years of reaction

| Thaumasite | a (Å) | b (Å) | c (Å) | α | β | γ | Vol (Å)$^3$ |
|---|---|---|---|---|---|---|---|
| **PDF 24-0038** | 11.04 | 11.04 | 10.39 | 90 | 90 | 120 | 1096.691 |
| **Refined Spectrum** | 11.0717 ± 0.0009 | 11.0717 ± 0.0009 | 10.4451 ± 0.0014 | 90 | 90 | 120 | 1108.84 ± 0.0020 |

## 4. Conclusions

The following three conclusions can be drawn from the experimental work described in the present paper:

1) The thaumasite synthesis in absence of ettringite demonstrates that this phase appears after 6 months. This means that it can precipitate directly from its components in solution. However, it must not be forgotten either that this is a laboratory experience –using pure reagents in very specific concentrations– and also that such conditions can hardly be found in real concretes.
2) Focusing on the *Ett series* allows us to see how thaumasite appears in the presence of ettringite after only a few days (both phases can be seen in the XRD spectrum after 7 days). That proves the catalyzing effect of ettringite.
3) Subsequently, ettringite disappears after 2 months in the *Ett series*, and the analytical results of samples obtained after 2 years through the refinement of thaumasite cell parameters would suggest that thaumasite is formed as a result of an ettringite transformation or conversion through the replacement of aluminum atoms with silicon atoms using a solid solution of both.

**Acknowledgements**

The authors wish to express their gratitude for the economic support received from the *Ministerio de Economía y Competitividad* [Spanish Ministry of Economy and Competitiveness] through the Research Project BIA-28798-C02-02 *"La reacción sulfática interna entre áridos con distintos compuestos de azufre y cementos de diferente contenido de aluminatos"* [Internal sulfate reaction between aggregates with different sulfur compounds and cements with different aluminate contents].

## References

[1] Bensted J 1999 Thaumasite - background and nature in deterioration of cements, mortars and concretes. *Cement & Concrete Composites* **21** 117-21

[2] Chinchón-Payá S, Aguado A, Nugteren H W and Chinchón S 2015 External sulfate attack in dam concretes with thaumasite formation. *Materiales de Construcción* **65**(317)

[3] Abubaker F, Lynsdale C and Cripps J 2014 Laboratory study of the long-term durability of buried concrete exposed to Lower Lias Clay. *Construction and Building Materials* **64** 130-40

[4] Hossack A M and Thomas M D A 2015 Varying fly ash and slag contents in Portland limestone cement mortars exposed to external sulfates. *Construction and Building Materials* **78** 333-41

[5] Aguilera J, Blanco-Varela M T, Martínez-Ramírez S, Palomo A, Riontino C, Sabbioni C, Toumbakari E E, Van Valen K and Zappia G 2001 Thaumasite formation in hydraulic mortars by atmospheric SO2 deposition. *Materiales de Construcción*

[6] Crammond N 2002 The occurrence of thaumasite in modern construction - a review. *Cement & Concrete Composites* **24**(3-4) 393-402

[7] The_UK_Government_Thaumasite_Expert_Group, *The thaumasite form of sulfate attack: Risks, diagnosis, remedial works and guidance on new construction*, in *Report of the Thaumasite Expert Group, Department of the Environment, Transport and the Regions: London*. 1999

[8] BRE_(Building_Research_Establishment). in *First International Conference on Thaumasite in Cementitious Materials*. 2002. BLDG RES ESTABLISHMENT, Watford, England

[9] Cement_&_Concrete_Composites 2003 Cement & Concrete Composites. **25**(8)

[10] Torres S M, Kirk C A, Lynsdale C J, Swamy R N and Sharp J H 2004 Thaumasite–ettringite solid solutions in degraded mortars. *Cement and Concrete Research* **34**(8) 1297-305

[11] Taylor H F W, *Cement chemistry*. 1997: Thomas Telford
[12] Crammond N 2003 The thaumasite form of sulfate attack in the UK. *Cement and Concrete Composites* **25**(8) 809-18
[13] Rahman M M and Bassuoni M T 2014 Thaumasite sulfate attack on concrete: Mechanisms, influential factors and mitigation. *Construction and Building Materials* **73** 652-62
[14] Aguilera J, Varela M T B and Vazquez T 2001 Procedure of synthesis of thaumasite. *Cement and Concrete Research* **31** 1163-68
[15] Kohler S, Heinz D and Urbonas L 2006 Effect of ettringite on thaumasite formation. *Cement and Concrete Research* **36**(4) 697-706
[16] Barnett S J, *X-ray Powder Diffraction Studies of Ettringite and Related Systems*. 1998: Staffordshire University
[17] Barnett S J, Macphee D E, Lachowski E E and Crammond N J 2002 XRD, EDX and IR analysis of solid solutions between thaumasite and ettringite. *Cement and Concrete Research* **32** 719-30
[18] Barnett S J, Macphee D E and Crammond N J 2003 Extent of immiscibility in the ettringite–thaumasite system. *Cement and Concrete Composites* **25**(8) 851-55
[19] Cody A M, Lee H, Cody R D and Spry P G 2004 The effects of chemical environment on the nucleation, growth, and stability of ettringite $[Ca_3Al(OH)_6]2(SO_4)_3 \cdot 26H_2O$. *Cement and Concrete Research* **34**(5) 869-81
[20] Rietveld H 1969 A profile refinement method for nuclear and magnetic structures. *Journal of Applied Crystallography* **2**(2) 65-71
[21] Young R A, *The Rietveld Method*. Monographs on Crystallography. 1995: Oxford University Press
[22] Rodríguez-Carvajal J, *FULLPROF: A Program for Rietveld Refinement and Pattern Matching Analysis*, in *Abstracts of the Satellite Meeting on Powder Diffraction of the XV Congress of the IUCr*. 1990: Toulouse (France). p. 127
[23] Hubbard C R, Evans E H and Smith D K 1976 The reference intensity ratio, I/Ic, for computer simulated powder patterns. *Journal of Applied Crystallography* **9**(2) 169-74
[24] Barnett S J, Adam C D and Jackson A R W 2000 Solid solutions between ettringite, $Ca_6Al_2(SO_4)_3(OH)_{12} \cdot 26H_2O$, and thaumasite, $Ca_3SiSO_4CO_3(OH)_6 \cdot 12H_2O$. *Journal of Materials Science* **35** (16) 4109 - 14

IOP Conf. Series: Materials Science and Engineering **897** (2020) 012003 doi:10.1088/1757-899X/897/1/012003

# Effects of process parameter variation on the surface roughness of polylactic acid (PLA) materials using design of experiments (DOE)

**C Burke, A Dalal, A Abukhalaf, and R Noorani**

Department of Mechanical Engineering, Loyola Marymount University, Los Angeles, CA 90045, USA.

Email: adalal@lion.lmu.edu; rinoorani@gmail.com

**Abstract.** 3D Printing (3DP) is an additive manufacturing technology used to rapidly build parts that are designed using 3D modeling software. 3DP builds a part by adding one layer of the working material at a time until the process is complete. One main concern with 3D printed samples is the high levels of surface roughness, which can result in the rejection of parts by many precision manufacturing companies. The objective of this research is to use the Design of Experiment (DOE) to analyze which factors influence the surface roughness of the part built from a 3D printer. In this research, a two-level, three-factor, full factorial design of experiment is used to select the best combination of factors that will minimize the surface roughness of parts made from Polylactic Acid (PLA) materials. The selected factors are printing orientation, nozzle diameter, and infill percentage. Based on the preliminary result, it is determined all the factors and their two-factor interactions are shown to significantly affect the surface roughness. However, it is shown that the nozzle diameter has had the most effect on surface roughness. These results will be explained in terms of the optical microscopy of the processed PLA test specimens.

## 1. Introduction

Fused deposition modeling (FDM) is a process that is used for fabricating solid prototypes from a computer-aided design (CAD) data file [1]. The process fabricates 3-D parts from a build-up of 2-D layers. In this process, Polylactic Acid (PLA) thermoplastic polymer is extruded through a heated nozzle to deposit the layers. In a previous paper, we showed that the orientation of the layers created anisotropic behavior in the tensile strength [2]. Other investigators have also experienced similar results [3].

Previous work has used the design of experiments (DOE) as a method to minimize surface roughness of the FDM- processed specimens of ABS polymers [4-5]. However, many of the factors used for influencing the surface roughness were entirely different in these studies. Also, these statistically designed experiments on FDM processing using PLA materials have not been physically interpreted in terms of the material properties and microstructure.

The purpose of this paper is to use the quality engineering tools to design, analyze and physically interpret our selection of the FDM processing factors and their levels for specifically PLA materials. It is expected that with DOE the best factor combination from the factors such as printing orientation, nozzle diameter, and infill percentage are experimentally determined to achieve the best surface roughness of the prototypes. There will be a total of eight possible growth combinations, in which the

IOP Conf. Series: Materials Science and Engineering **897** (2020) 012003 doi:10.1088/1757-899X/897/1/012003

surface roughness produced will be measured. Thereafter, using DOE, a determination can be made of what level each factor should be to consistently get a smoother surface.

## 2. Material properties

PLA is one of the most common thermoplastic materials used in 3DP because it is environmentally friendly and is a biodegradable polymer created from sugar plants such as tapioca, corn, and sugarcane. It also does not require a heated build plate because PLA melts at a very low temperature. Its melting point is at around 160°C, but bonds better at 180° [6]. PLA is also much more brittle than other thermoplastics. However, some improvements to it are being made in order to increase its flexibility and reduce carbon footprint during the creation process. One of the reasons why it is so popular is because of its ease of creation from whatever natural sugar plants are locally available.

## 3. Experimental methods

The 12-step Taguchi process was used for our design of experiments (DOE) [7]. First, the problem of concern was the high surface roughness of FDM-processed PLA test specimens. Second, our objective was to minimize surface roughness. Third, the surface roughness was selected as the quality characteristics, i.e., the response. Fourth, the factors were determined by brainstorming and were recorded on a cause and effect diagram [8]. Fifth, three factors and no noise factors were selected. Sixth, a two-level experiment was selected.

Seventh, a $2^3$ orthogonal array was used for the DOE. Eighth, all the selected factors could potentially interact with each other. Ninth, all the interactions were listed. Tenth, the trials were randomized and four specimens per trial were used. Eleventh, the column effects method and plots of the response versus the effects of factors and interactions were used to analyze the data. Twelfth, the 95% statistically significant factors and interactions were identified. Our results were interpreted in terms of surface roughness and microstructure to validate our results.

The FDM 3D printer used to fabricate the test specimens was the Tevo Tarantula [9]. This is a popular choice of the 3D printer due to its low cost and similarity to other more expensive options such as the Prusa i3. While it lacks the user-friendly features such as automatic bed leveling and filament sensors, the core components are nearly identical to many FDM printers: aluminum frame utilizing 3D printed brackets, belt-driven movement for the bed platform and extruder carriage, threaded rod actuation for the z-axis, use of an Arduino-based motherboard and g-code file format, etc. The filament material used was a 1.75mm black PLA from Hatchbox [10]. The filament is fed into the heated nozzle, located on the carriage, from a Bowden extruder that is mounted to the frame. This material is melted and extruded out of a brass nozzle to form a molten polymer. The polymer was deposited at locations on the X-Y plane according to the movement dictated by the g-code commands. Then the head is moved vertically in the Z-axis to deposit a new layer on top of the previous one. In this way, the 3-D solid model was built-up by multiple depositions of 2D layers.

Process parameters of interest such as layer thickness and infill density were modified using the 3D printing slicer software Ultimaker CURA 4.1. While changing the diameter of filament extruded can be accomplished solely through the slicer settings, a more extreme difference in filament size can only be accomplished by swapping between two brass nozzles with different internal diameters; in this case, a 0.2 mm nozzle was tested along with a 1.0 mm nozzle. Each layer is formed by first depositing the perimeter (outside contour) around the X-Y plane of the test specimen design and then filling the inside of the perimeter with a raster pattern that had a determined orientation based on the settings used in Cura. The layer is partially bonded to the underlying structure. This process was like the 2D lay-up of 0°/90° and ±45°laminate composites.

The geometry of the test specimens was made using Solidworks 2019 CAD software and consists of a 25.4mm x 76.2mm x 6.35mm rectangular cuboid. This model is then exported as a STL file and loaded into Cura where the g-code that the 3D printer can read is generated based on the selected process parameters. After all the samples were printed, the surface roughness of the 25.4mm x 76.2mm side was measured with the Taylor Hobson Surtronic 25 profile measurement system.

IOP Conf. Series: Materials Science and Engineering **897** (2020) 012003 doi:10.1088/1757-899X/897/1/012003

## 4. Design of Experiments

A cause-effect diagram was used to list the possible causes affecting the surface roughness of the test specimens. Some of the possible causes were the build specifications (road width, layer thickness, raster orientation), machine environment (model temperature), PLA materials (density). Three factors were selected for this experiment: (A) printing orientation, (B) nozzle diameter and (C) infill percentages at two levels, i.e., low (-1) and high (+1), respectively in Table 1. The three factors had a potential effect on the surface roughness of the samples. Infill percentage is the amount of material that is filled in the sample, the sample can be hollow or can be filled with the material up to 100%. The strength of the sample depends on the infill percentage. The second factor is print orientation, the sample can be printed in two orientations, flat or standing. Print orientation decides how it be printed and how much support material would be used. The third factor is nozzle diameter, which affects most in printing the samples. Higher the nozzle diameter, the higher the surface roughness, so the lowest surface roughness can be obtained when the sample is printed with low nozzle diameter.

A ($2^3$) orthogonal design matrix was utilized as shown in Table 2. Eight experimental trials were conducted which were randomly selected to the run numbers in Table 2. Four replications were conducted for each run.

**Table 1.** Factors and levels

| Factors | Parameters | Low (-1) | High (1) |
|---|---|---|---|
| A | infill percentage | 5% | 100% |
| B | orientation | flat | standing |
| C | nozzle diameter | 0.2 mm | 1.0 mm |

**Table 2.** Orthogonal experimental design matrix

| Run # | Factors | | |
|---|---|---|---|
| | A | B | C |
| 4 | -1 | -1 | -1 |
| 5 | -1 | -1 | 1 |
| 3 | -1 | 1 | -1 |
| 8 | -1 | 1 | 1 |
| 7 | 1 | -1 | -1 |
| 6 | 1 | -1 | 1 |
| 1 | 1 | 1 | -1 |
| 2 | 1 | 1 | 1 |

The relevant equations used to find the system equation in the DOE analysis are shown below:

$$\overline{\sigma^2} = \Sigma(\sigma_i^2)/2^k \quad (1)$$

In Eq. (1), $\overline{\sigma^2}$ is the pooled variance estimate, $\sigma_i$ is the individual standard deviation for the factor and k is the number of factors (3). This equation is used to determine the standard error (se), which is then used to formulate the 95% confidence interval in the DOE plot.

$$se = \sqrt{\frac{\overline{\sigma^2}}{n \cdot 2^{k-2}}} \quad (2)$$

In Eq. (2), $\overline{\sigma^2}$ is the pooled variance estimate, $N = n \cdot 2^k$ measurements, k is the number of factors and n is the number of replications. This equation finds the standard errors. The standard errors are then used to locate the 95% confidence envelope seen in the DOE plot by taking $\pm 2 \cdot se$. The regression model is given by following equation.

$$Y = \beta_0 + \sum_i \beta_i x_i + \sum_i \sum_j \beta_{ij} x_i \cdot x_j \tag{3}$$

In Eq. (3), $\beta_o$ is the grand mean of the regression equation and $\beta_i$ is the coefficients in the regression equation for each of the factors that were found to have a real effect on the system. These coefficients are found by dividing the effect by 2. $x_i$ and $x_j$ are the corresponding high (1) or low (-1) values of the factors with real effects.

This DOE technique has been implemented and proven effective throughout the industry for optimizing processes. It is hypothesized that this DOE technique can be used to effectively reduce the surface roughness of prototypes grown with a Tevo Tarantula 3d printer.

## 5. Results

The measured surface roughness, in $\mu m$, for each of the 32 samples and the average surface roughness of each run is shown in Table 3.

**Table 3.** Surface roughness (μm) data for the design matrix

| Run # | $R_{a1}$(μm) | $R_{a2}$ (μm) | $R_{a3}$ (μm) | $R_{a4}$ (μm) | Avg. $R_a$ (μm) | Variance |
|---|---|---|---|---|---|---|
| 4 | 8 | 9.2 | 14.2 | 10.6 | 12.4 | 7.29 |
| 5 | 26.2 | 33.6 | 40.0 | 28.8 | 34.4 | 37.21 |
| 3 | 14.8 | 15.2 | 16.0 | 15.2 | 15.3 | 0.25 |
| 8 | 44.0 | 44.0 | 45.0 | 43.8 | 44.2 | 0.25 |
| 7 | 16.6 | 12.4 | 12.8 | 14.0 | 14.0 | 3.61 |
| 6 | 28.2 | 24.4 | 24.4 | 23.2 | 25.1 | 4.84 |
| 1 | 14.8 | 15.0 | 14.6 | 20.8 | 16.3 | 9 |
| 2 | 42.2 | 43.2 | 41.8 | 41.0 | 42.1 | 0.81 |

It is clear from Table 3 that the surface roughness varies from a minimum of 12.4 $\mu m$ of Run # 4 to a maximum of 44.2 $\mu m$ of Run # 8. It represents a decrease in surface roughness by 250%. The response table for computing the effect of each factor and each possible factor interaction is shown in Table 4.

**Table 4.** Response table

| Random Order | Std. Order | Response | A | | B | | C | | AB | | AC | | BC | | ABC | |
|---|---|---|---|---|---|---|---|---|---|---|---|---|---|---|---|---|
| | | | -1 | 1 | -1 | 1 | -1 | 1 | -1 | 1 | -1 | 1 | -1 | 1 | -1 | 1 |
| | | | 1 | 2 | 1 | 2 | 1 | 2 | 1 | 2 | 1 | 2 | 1 | 2 | 1 | 2 |
| 4 | 1 | 10.5 | 10.5 | | 10.5 | | 10.5 | | | 10.5 | | 10.5 | | 10.5 | 10.5 | |
| 5 | 2 | 32.2 | 32.2 | | 32.2 | | | 32.2 | | 32.2 | 32.2 | | 32.2 | | | 32.2 |
| 3 | 3 | 15.3 | 15.3 | | | 15.3 | 15.3 | | 15.3 | | | 15.3 | 15.3 | | | 15.3 |
| 8 | 4 | 44.2 | 44.2 | | | 44.2 | | 44.2 | 44.2 | | 44.2 | | | 44.2 | 44.2 | |
| 7 | 5 | 14 | | 14 | 14 | | 14 | | 14 | | 14 | | | 14 | | 14 |
| 6 | 6 | 25.1 | | 25.1 | 25.1 | | | 25.1 | 25.1 | | | 25.1 | 25.1 | | 25.1 | |
| 1 | 7 | 16.3 | | 16.3 | | 16.3 | 16.3 | | | 16.3 | 16.3 | | 16.3 | | 16.3 | |
| 2 | 8 | 42.1 | | 42.1 | | 42.1 | | 42.1 | | 42.1 | | 42.1 | | 42.1 | | 42.1 |
| Total | | 199.7 | 102.2 | 97.5 | 81.8 | 117.9 | 56.1 | 143.6 | 98.6 | 92.1 | 106.7 | 93 | 88.9 | 110.8 | 96.1 | 103.6 |
| #of values | | 8 | 4 | 4 | 4 | 4 | 4 | 4 | 4 | 4 | 4 | 4 | 4 | 4 | 4 | 4 |
| Average | | 24.9625 | 25.55 | 24.375 | 20.45 | 29.475 | 14.025 | 35.9 | 24.65 | 23.025 | 26.675 | 23.25 | 22.225 | 27.7 | 24.025 | 25.9 |
| Effects | | | -1.175 | | 9.025 | | 21.875 | | -1.625 | | -3.425 | | 5.475 | | 1.875 | |

IOP Conf. Series: Materials Science and Engineering **897** (2020) 012003 doi:10.1088/1757-899X/897/1/012003

The response graph for surface roughness (in micrometer) is shown in Figure 1 with a grand average of 24.965 (μm) by evaluating Table 4. The standard error (se) was computed to be 4.93 (μm). The upper bound is the grand average plus 2se which is 34.8225 (μm) and the lower bound is the grand average minus 2se which is 15.1025 (μm).

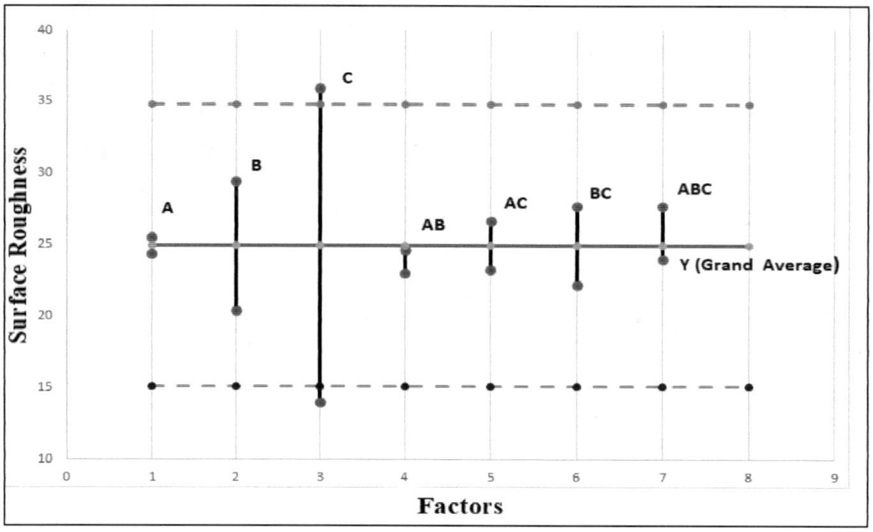


**Figure 1.** DOE mean plot for average surface roughness data

The interaction diagram for factor A (infill percentage) and B (orientation) is shown in Figure 2.

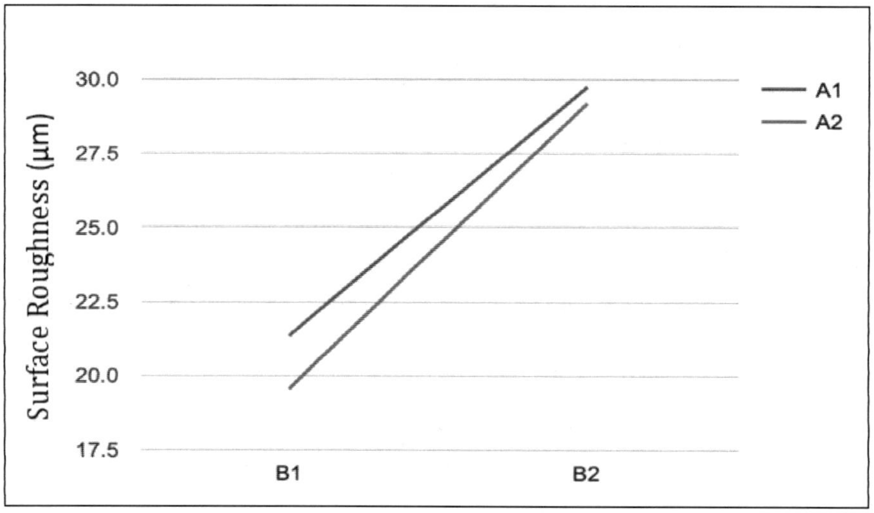


**Figure 2.** Interaction diagram for factors A and B

The interaction diagram for factor B (orientation) and C (nozzle diameter) is shown in Figure 3.

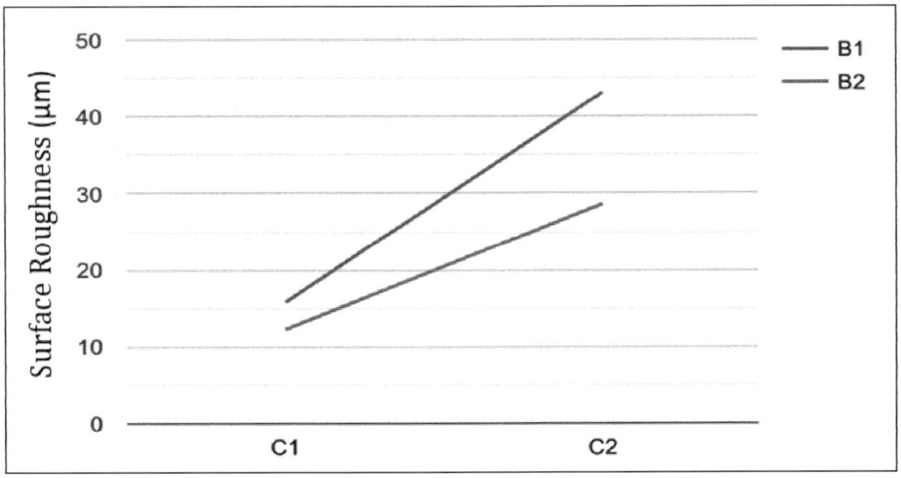


**Figure 3.** Interaction diagram for factors B and C

The interaction diagram for factor A (infill percentage) and C (nozzle diameter) is shown in Figure 4.

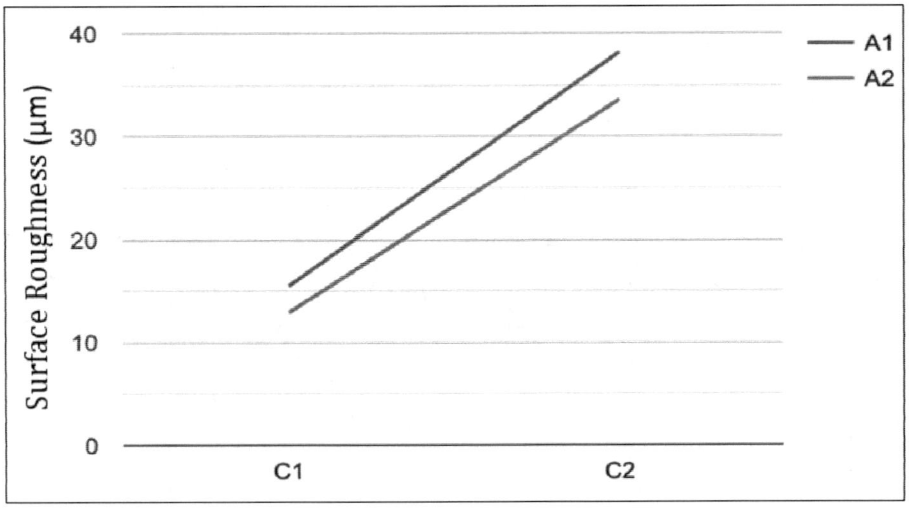


**Figure 4.** Interaction diagram for factors A and C

The calculations of variance, standard error, and $Y_{min}$ is shown below:

$$\overline{\sigma^2} = \frac{\Sigma(\sigma_i^2)}{2^k} = \frac{1555.967}{6} = 194.4458$$

$$se = \sqrt{\frac{\overline{\sigma^2}}{n \cdot 2^{k-2}}} = \sqrt{\frac{194.4958}{8}} = 4.93$$

$$Y = \beta_0 + \sum_i \beta_i x_i + \sum_i \sum_j \beta_{ij} x_i \cdot x_j = 24.9625 + \left(\frac{21.895}{2}\right).(-1) = 14.025$$

$$Y_{min} = 14.025 \text{ µm}$$

IOP Conf. Series: Materials Science and Engineering **897** (2020) 012003 doi:10.1088/1757-899X/897/1/012003

The $Y_{min}$ from the system equation validates the minimum surface roughness in Run # 4 in Table 3.

## 6. Physical Interpretation

The selected process factors must physically alter the surface structure of the 3D printed samples to vary the measured surface roughness values between the eight specimens. Any variation in surface roughness is the result of the different factors used, and these changing factors are also visible in the optical microscopy of the specimens. Per the results obtained, build orientation (flat or standing) and extruder nozzle diameter (0.2 or 1.0 mm) had the largest impact on surface roughness and the most obvious impact on the physical appearance of the samples. While the amount of infill used can impact the surface quality of some 3D printed models [1], for the combination of process parameters used in this study, there was no statistically significant impact on surface roughness due to infill percentage. However, it is hypothesized that lowering the number of layers above the infill (this setting is called "Top Layers" in Cura) would result in infill percentage having more of an effect on surface roughness as there would be fewer layers bridging the gap between infill lines and smoothing out the surface.

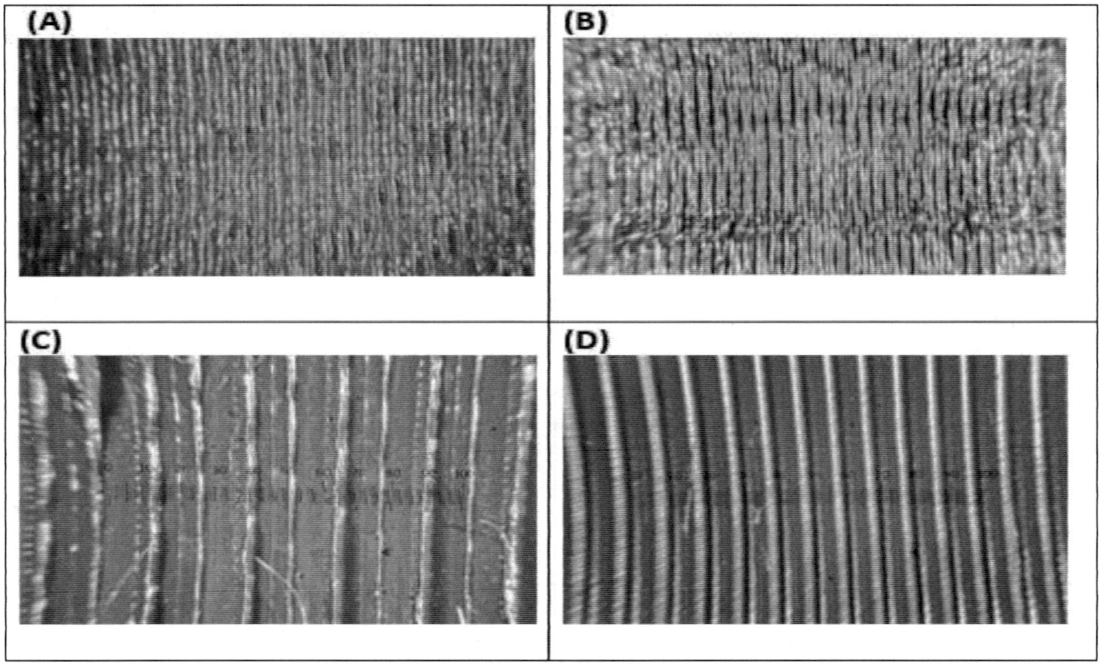


**Figure 5.** Physical interpretation

When looking for the physical interpretation of these results through digital microscopy, the differences in surface roughness values become evident. Figure 5 shows the pictures obtained by digital microscopy of four specimen surfaces where the only difference between the sets is to build orientation and extruder nozzle diameter. Pictures (A) and (B) show the samples that were printed using a 0.2mm nozzle where pictures (C) and (D) are from samples printed with a 1.0mm nozzle. Similarly, (A) and (C) are from the samples printed flat on the build plate and (B) and (D) are from samples printed standing vertically on the build plate.

To show the physical effects of build orientation on surface roughness, Figure 5 (A) and (C) picture the horizontal surface of a single layer where the nozzle both deposits the material and critically irons out any bumpy sections, creating a smoother, more uniform plane. Pictures (B) and (D) on the other hand show the layer lines as the model is built vertically. Due to the way that FDM printers must not only deposit material but also ensure that the deposited plastic fuses with the layer below it, the layers must be squeezed together with the force which causes the edges of the layers to bulge outwards as the molten material is compressed by the nozzle into the solid layer beneath. This manifests in Figure 5 as dark shadows in the pictures as the bulging layer lines block light from entering the valleys between the layers.

Now looking into the physical impact of the two nozzle sizes, the extruded lines of plastic in (A) and (B) are quite small, around 0.2mm wide in both cases and exhibit mild waviness throughout the model. It can be seen in these photos that the defects of a single line are contained within a small area and thus provide a small amount of surface roughness. Essentially the printer is extruding these smaller lines closer together which gives it less room for error. Conversely, in pictures (C) and (D), the gaps between solid layers and lines are larger and show more extreme bumps in the form of large dark shadows. Here the larger nozzle exacerbates the natural defects inherent in all FDM 3D printing by extruding a larger layer or line of plastic. One of these defects present here includes hair-like strings of plastic caused by the larger nozzle knocking into the neighboring lines and pulling the plastic into strings. This is evident in Figure 5 and confirmed by the results obtained for surface roughness for each of the samples.

## 7. Discussion

Since all factors in the response graph except C (nozzle diameter) (Figure 1) are within ±2se from the grand average we can conclude that nozzle diameter has a significant effect on surface roughness of 3DP samples. Since nozzle diameter showed a significant effect on the surface roughness, further a regression equation is useful to predict surface roughness. It is seen in the response graph (Figure 1) above that factor C, nozzle diameter, has the largest range.

Since there is no intersection seen in the interaction diagrams (Figures 2 – 4) it is concluded that there were no real effects to surface roughness due to the interaction of the chosen three factors.

## 8. Conclusion

After completing this experimental study, the following conclusions can be made:

- The hypothesis that the DOE technique can effectively be used to minimize the surface roughness of RP prototypes is accepted.
- The lowest surface roughness of 12.4 μm is achieved in Run # 4, which uses low (5%) infill percentage, flat orientation and nozzle diameter (0.2 mm).
- Using statistical analysis including interaction diagrams and a response graph, it is shown that from infill percentage, print orientation, and nozzle diameter, nozzle diameter has the most impact on the surface roughness.
- The interaction graphs show that all factors are independent of each other.
- The experiment can be repeated with different factors and different levels by replicating more samples per run.

## Reference

[1] Stratasys, Fused Deposition modeling, website: http:/www.stratasys.com

[2] Es-said, O, Noorani R, Mendelson, M, Foyos, J, and Marloth, R 2000 "*Effects of Layer Orientation on Mechanical Properties of Rapid Prototyped Samples*," Materials and Manufacturing Processes, 15, 107-122

[3] Kanaoka Y, Kazi P, Petrosian S, James C, Mendelson M, and Noorani R, "*Design of Experiments for the Fused Deposition Modeling Process*

[4] Horvath D, Noorani R, Mendelson M 2007 "*Improvement of Surface Roughness on ABS 400 Polymer Materials Using Design of Experiments (DOE)*". Materials Science Forum, Vol, 561-565 pp. 2389-2392

[5] Ayivi V, Concialdi J, Villanueva J, Alzebaq R, Noorani R 2016 "*Investigating the Surface roughness of Rapid Prototyped Samples using Design of Experiments.*" Presented and published at the ICLIST conference in Pattaya, Thailand

[6] Noorani R 2006 *Rapid Prototyping- Principles and Applications*, John Wiley & Sons, Hoboken, NJ
[7] Ross, PJ 1996, *Taguchi Techniques for Quality Engineering*, 2nd Edition, McGraw-Hill
[8] Summers, D.C. 1997 Quality, Prentise-Hall, Upper Saddle River, NJ
[9] Tevo tarantula, website: https://tevo3dprinterstore.com
[10] https://www.hatchbox3d.com/products/3d-pla-1kg1-75-bl

IOP Conf. Series: Materials Science and Engineering **897** (2020) 012004 doi:10.1088/1757-899X/897/1/012004

# Does the degree of tensile strain have an impact on the cracking behavior of vertical structural elements?

**T Chrysanidis[1*] and V Panoskaltsis[1]**

[1]Department of Civil Engineering, Demokritos University of Thrace, Xanthi, Greece

[*]Email: thchrysa@civil.duth.gr

**Abstract.** Given the random nature of crack formation, research into reinforced concrete members in the context of cracking behavior proves difficult. Therefore, widely accepted methodologies for predicting crack characteristics, e.g. crack width and spacing and number of cracks, have not been developed yet. Furthermore, cracking for members strained to high degrees of elongation, as happens during earthquakes, has not been investigated before. This experimental work aims to look into the impact, primarily of tensile deformation's mechanical factor, in terms of cracking behaviour. Four test specimens were strained under uniaxial tensile loading. The degrees of elongation used were equal to 10‰, 20‰, 30‰ and 50‰. Useful conclusions concerning cracking behaviour are derived.

## 1. Introduction

Several international researchers have explored the phenomenon of cracking in reinforced concrete structures [1]–[23]. Although most of the given structures are typically reinforced in two directions, most studies carried out worldwide to investigate the behaviour of cracking have involved uniaxially tensiled reinforced concrete members with reinforcement in only one direction. To date there is not yet a widely established and accepted methodology for predicting cracking characteristics, e.g. crack widths and spacings between cracks [7], [9], [10], [23]–[26]. In addition, most of the existing research conducted has been strictly limited to the state of the stabilized crack pattern only and does not involve cracking behaviour and crack characteristics deep in the yield region [27], [28].

In the framework of the current study, an experimental program has been conducted involving reinforced concrete members detailed in two directions using longitudinal rebars and transverse reinforcement in the form of ties. This is a common construction practice used in the vast majority of concrete structures. As per the results outlined within this work, cracking behaviour and the crack characteristics are discussed. Useful conclusions concerning cracking under different degrees of elongation are derived. The experimental procedure of the current study has taken place in Aristotle University of Thessaloniki at the Laboratory of Strength of Materials but the result analysis has taken place in Democritus University of Thrace.

## 2. Experimental program

### *2.1. Detailing of test specimens*

The current experimental program consists of four test specimens. The thickness of each specimen is 7.5 cm and the length of the cross-section is 15 cm. The ratio between the length and the thickness of the cross-sectional area is equal to 2, which is a typical ratio for constructing reinforced concrete columns. The total height of the test specimen is equal to 90 cm. Each of the four specimens is subjected to a uniaxial central tensile loading. The main test element is between the metal plates and its height is 64 cm (Fig. 1).

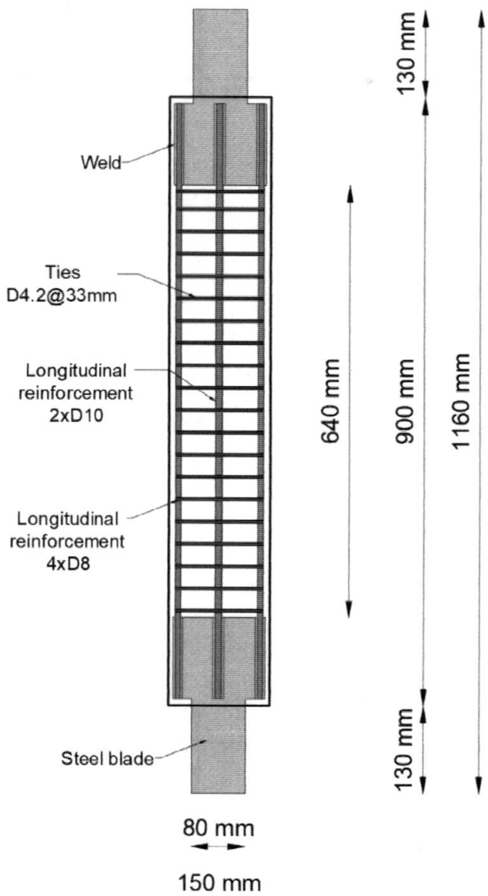


**Figure 1.** Vertical reinforcement layout for specimens

Table 1 shows the geometrical and detailing characteristics of all four specimens. All four segments tested here have been detailed in two directions through deformed bars in terms of reinforcement. The reinforcement of each specimen simulates a typical reinforcement found in the reinforced concrete columns of typical construction buildings or in the confined boundaries of reinforced concrete seismic walls. The construction scale used to simulate typical columns or typical confined boundaries was equal to 1:3, which is commonly used for research purposes worldwide [29], [30].

The number of longitudinal bars is equal to six. Four of them have a diameter of 8 mm and two of them have a diameter of 10 mm. The longitudinal reinforcement ratio is equal to 3.19%. The transverse reinforcement consists of transverse ties placed along the height of the prism. The centre-to-centre distance between two ties is about 3.3 cm and the diameter of each tie is 4.2 mm. The only variable differentiating specimens from each other is the tensile strain. The nominal tensile strain takes values equal to 10.00%, 20.00%, 30.00% and 50.00%. It is well known that in real constructions, tensile strains up to 30.00% have been observed [31], [32]. Also, modern seismic and concrete codes

IOP Conf. Series: Materials Science and Engineering **897** (2020) 012004 doi:10.1088/1757-899X/897/1/012004

have related provisions allowing large tensile strains for reinforcement bars [33]–[39]. The name of each specimen is of the type H-"Number", where "H" corresponds to the high longitudinal reinforcement ratio of the elements and the number shows the degree of elongation applied to the specimen in question.

**Table 1.** Geometrical and detailed properties of element specimens.

| N/A | Specime n name | Length (cm) | Thicknes s (cm) | Effective height (cm) | Longitudinal reinforceme nt | Longitudinal reinforceme nt ratio $[\rho_l]$ (%) | Transverse reinforceme nt | Nominal tensile strain (‰) |
|---|---|---|---|---|---|---|---|---|
| 1 | H-10 | 15 | 7.5 | 64 | 4xD8 + 2xD10 | 3.19 | D4.2@33 mm | 10.00 |
| 2 | H-20 | 15 | 7.5 | 64 | 4xD8 + 2xD10 | 3.19 | D4.2@33 mm | 20.00 |
| 3 | H-30 | 15 | 7.5 | 64 | 4xD8 + 2xD10 | 3.19 | D4.2@33 mm | 30.00 |
| 4 | H-50 | 15 | 7.5 | 64 | 4xD8 + 2xD10 | 3.19 | D4.2@33 mm | 50.00 |

### *2.2. Test setup for loading*

A universal testing machine was used to apply the load. For example, for specimen H-20, the nominal degree of elongation is 20.00‰. Fig. 2 displays the experimental configuration for imposing the tensile load. The rate of loading was slow, of the order of 4 mm/min, so no result was affected by the influence of the strain rate. This is well established in international bibliography and it is a common practice [40].

**Figure 2.** Loading test configuration.

## 3. Discussion of results

### *3.1. Experimental findings*

Fig. 3 presents the change of tensile strain relative to the tensile load imposed on the test elements. It is obvious that specimens with a high nominal degree of elongation are subjected to high tensile strains too and vice versa. It is noted that the nominal degrees of tensile strain (10‰, 20‰, 30‰, 50‰)

IOP Conf. Series: Materials Science and Engineering **897** (2020) 012004 doi:10.1088/1757-899X/897/1/012004

are slightly different from the residual actual tensile degrees, but these differences are small and negligible.

After conducting the experiments, different cracking formations and eventually cracking characteristics were noticed for each specimen. Fig. 4 shows the state of each specimen after the end of the uniaxial tensile test. Cracks of small width are obvious for specimens with low degrees of tensile strain (10‰ and 20‰), while cracks of moderate and large width are present for specimens strained under larger degrees of elongation (30‰ and 50‰). It is apparent that the final cracking formation differs between the specimens, depending on the tensile strain they have sustained.

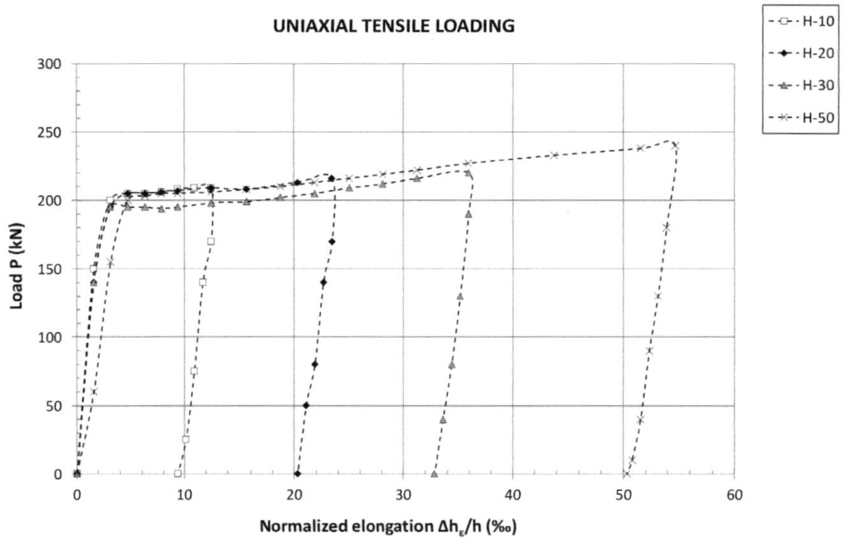


**Figure 3.** Diagram of tensile load versus elongation.

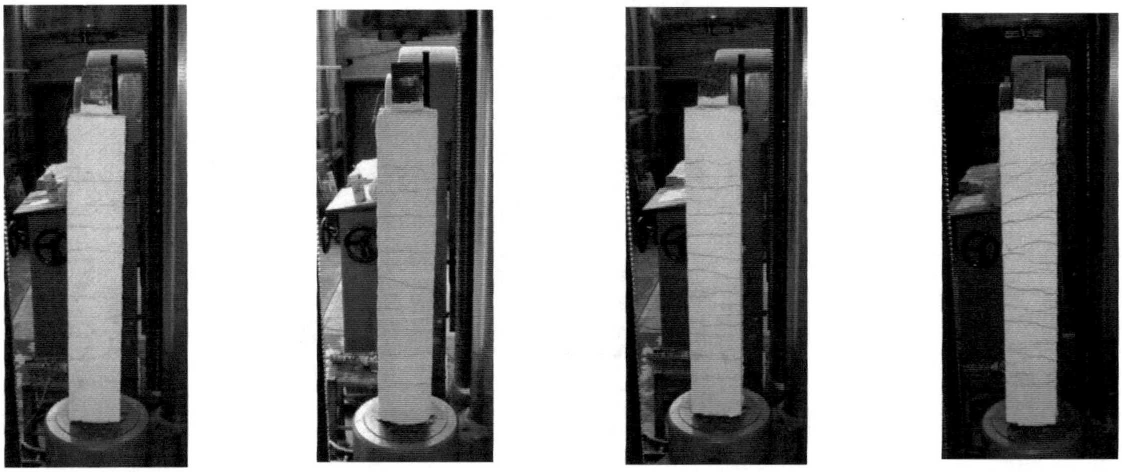

(a) (b) (c) (d)

**Figure 4.** Specimens after the uniaxial tensile test: (a) H-10, (b) H-20, (c) H-30, (d) H-50.

### *3.2. Analysis of experimental findings*

The results of the analysis of the test findings for all segments are brought together in the following tables and diagrams. Table 2 and Table 3 present the width and spacing characteristics of the cracks. Fig. 5 displays the variation of the crack width and spacing characteristics. Fig. 7 – Fig. 13 use bar charts to display the results for the same type of cracking characteristics, e.g. minimum width, average spacing, etc.

**Table 2.** Crack width characteristics.

| N/A | Specimen | Number of cracks [N] | Minimum crack width [$W_{min}$] (mm) | Maximum crack width [$W_{max}$] (mm) | Average crack width [$W_{ave}$] (mm) | $W_{min}/W_{ave}$ | $W_{max}/W_{ave}$ | $W_{max}/W_{min}$ |
|---|---|---|---|---|---|---|---|---|
| 1 | H-10 | 11 | 0.2 | 1.0 | 0.655 | 0.31 | 1.53 | 5.00 |
| 2 | H-20 | 12 | 0.2 | 1.5 | 1.050 | 0.19 | 1.43 | 7.50 |
| 3 | H-30 | 13 | 0.6 | 2.0 | 1.354 | 0.44 | 1.48 | 3.33 |
| 4 | H-50 | 13 | 0.8 | 3.2 | 2.277 | 0.35 | 1.41 | 4.00 |

**Table 3.** Crack spacing characteristics.

| N/A | Specimen | Number of cracks [N] | Minimum crack spacing [$W_{min}$] (cm) | Maximum crack spacing [$W_{max}$] (cm) | Average crack spacing [$W_{ave}$] (cm) | $S_{min}/S_{ave}$ | $S_{max}/S_{ave}$ | $S_{max}/S_{min}$ |
|---|---|---|---|---|---|---|---|---|
| 1 | H-10 | 11 | 4.3 | 10.4 | 6.540 | 0.66 | 1.59 | 2.42 |
| 2 | H-20 | 12 | 2.6 | 10.6 | 5.855 | 0.44 | 1.81 | 4.08 |
| 3 | H-30 | 13 | 2.6 | 7.6 | 5.533 | 0.47 | 1.37 | 2.92 |
| 4 | H-50 | 13 | 2.9 | 9.0 | 5.575 | 0.52 | 1.61 | 3.10 |

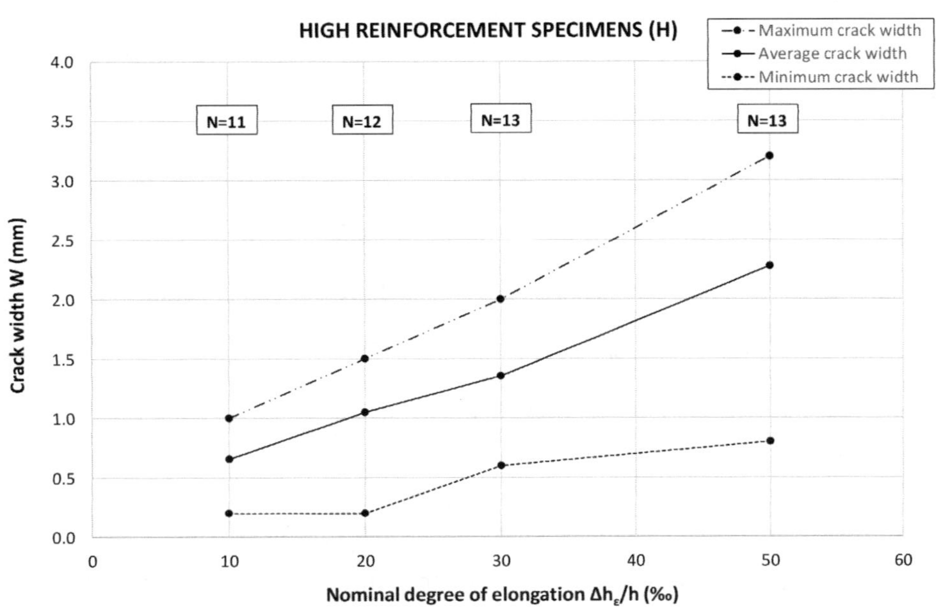


**Figure 5.** Change of crack width relative to the degree of elongation.

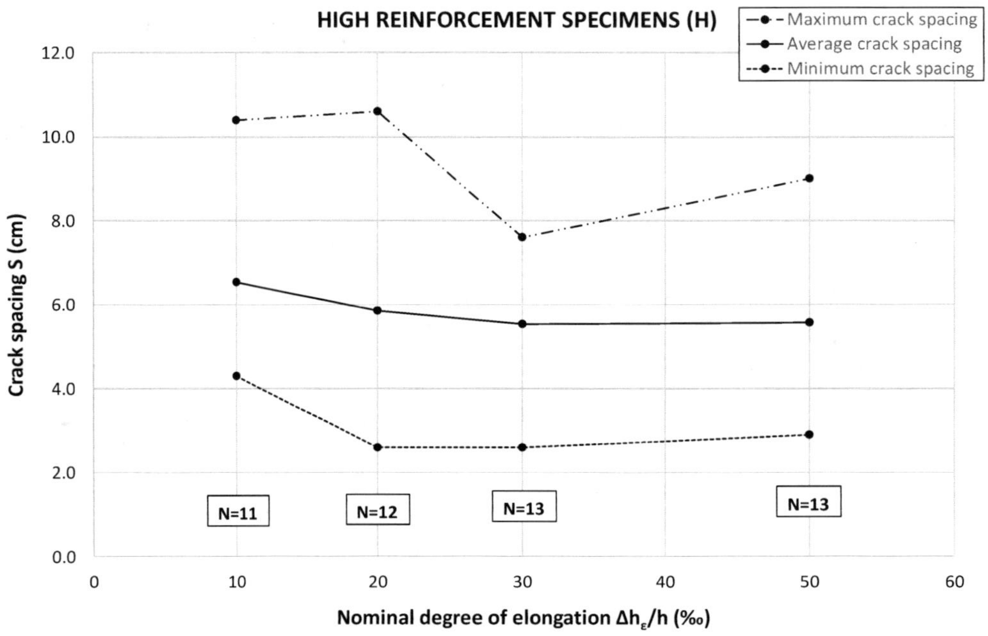


**Figure 6.** Change of crack spacing relative to the degree of elongation.

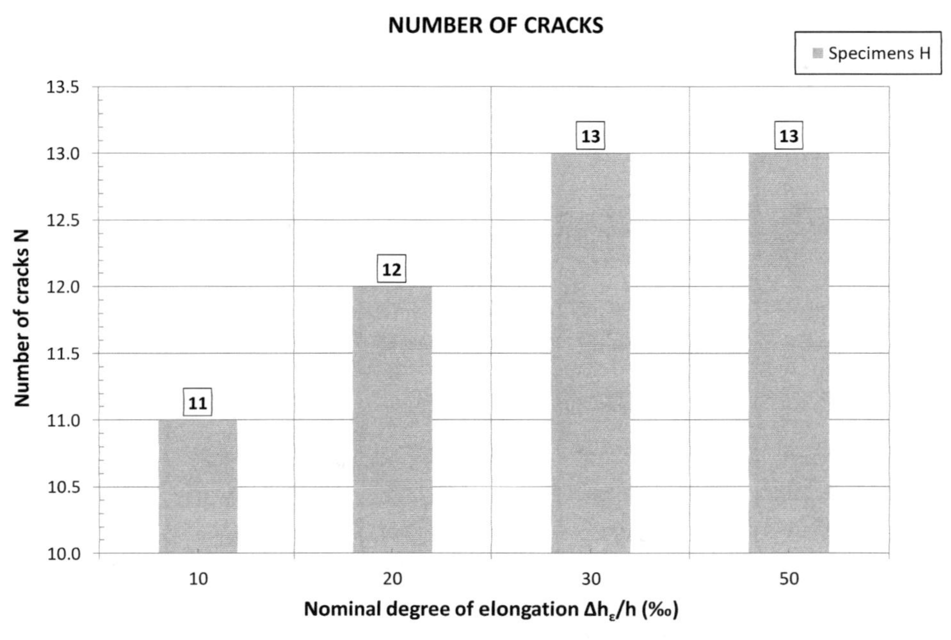


**Figure 7.** Bar chart of number of cracks regarding the degree of elongation.

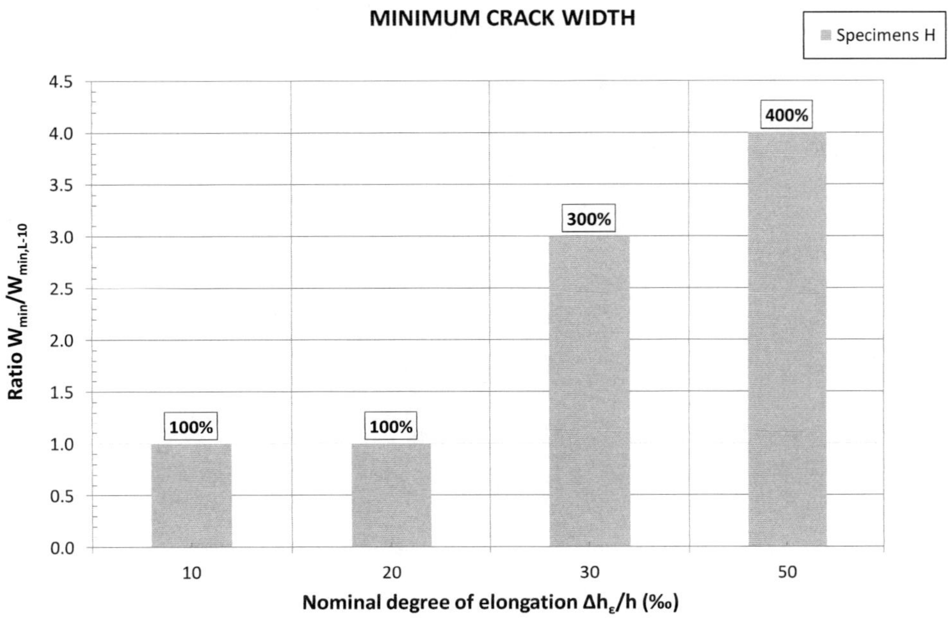


**Figure 8.** Bar chart of minimum crack width as a percentage of the minimum crack width of the reference specimen.

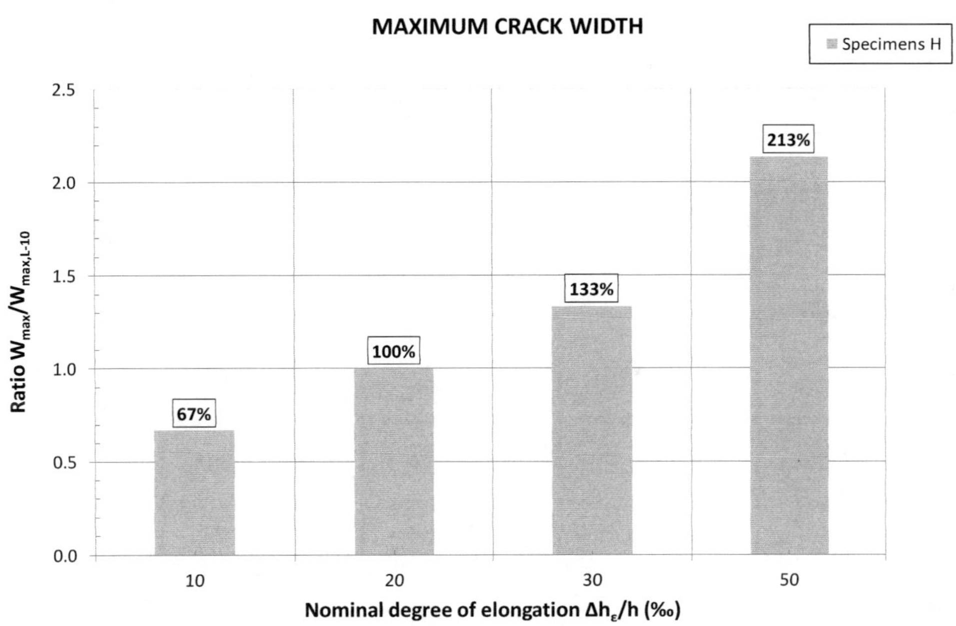


**Figure 9.** Bar chart of maximum crack width as a percentage of the maximum crack width of the reference.

IOP Conf. Series: Materials Science and Engineering **897** (2020) 012004 doi:10.1088/1757-899X/897/1/012004

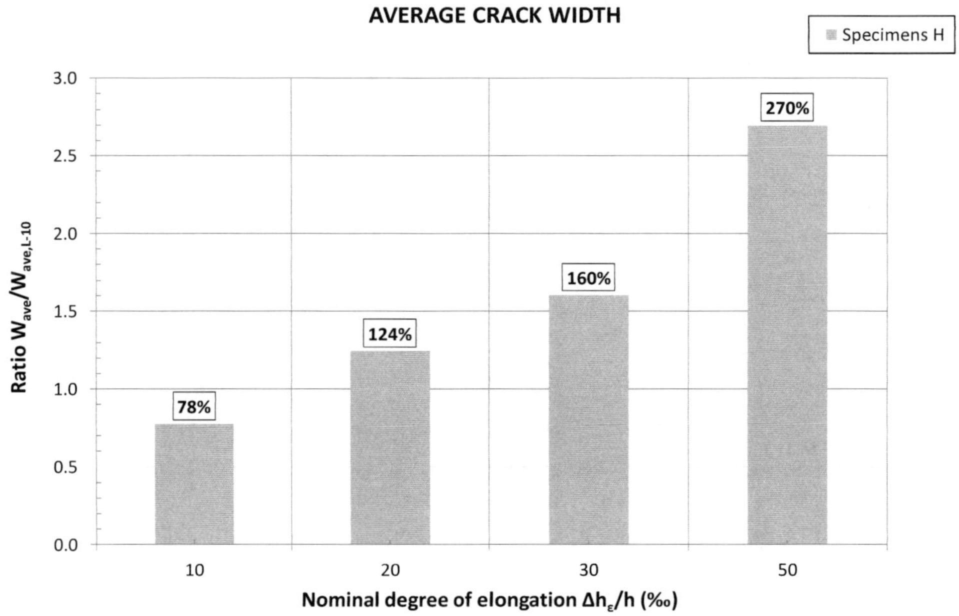


**Figure 10.** Bar chart of average crack width as a percentage of the average crack width of the reference specimen.

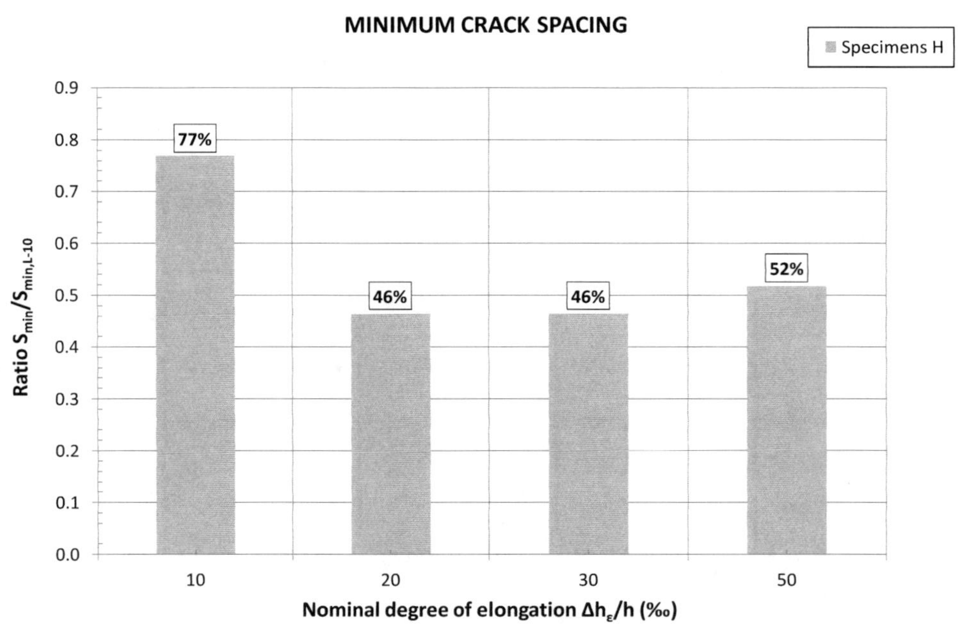


**Figure 11.** Bar chart of minimum crack spacing as a percentage of the minimum crack spacing of the reference specimen.

IOP Conf. Series: Materials Science and Engineering **897** (2020) 012004 doi:10.1088/1757-899X/897/1/012004

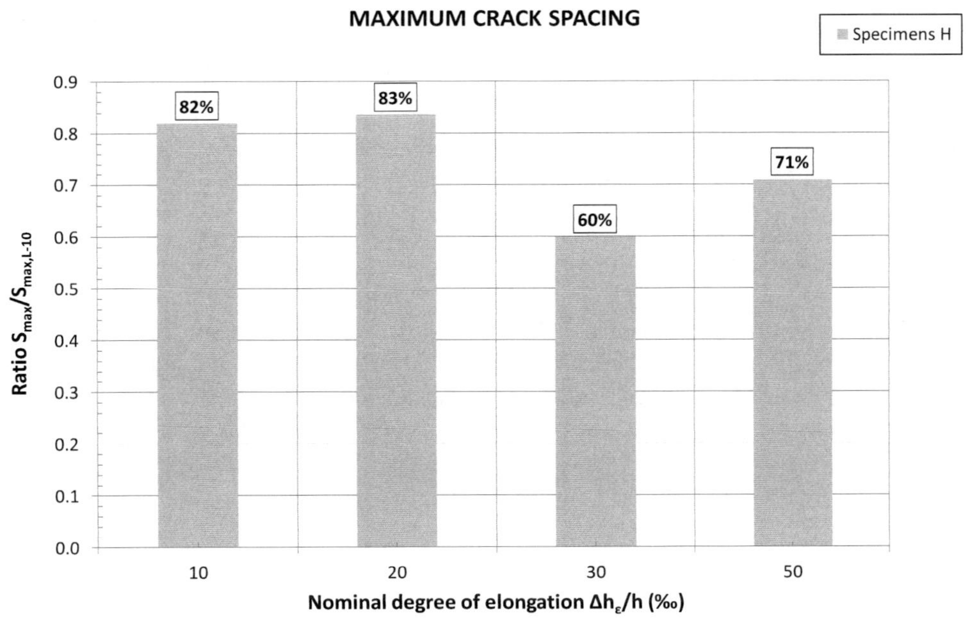


**Figure 12.** Bar chart of maximum crack spacing as a percentage of the maximum crack spacing of the reference specimen.

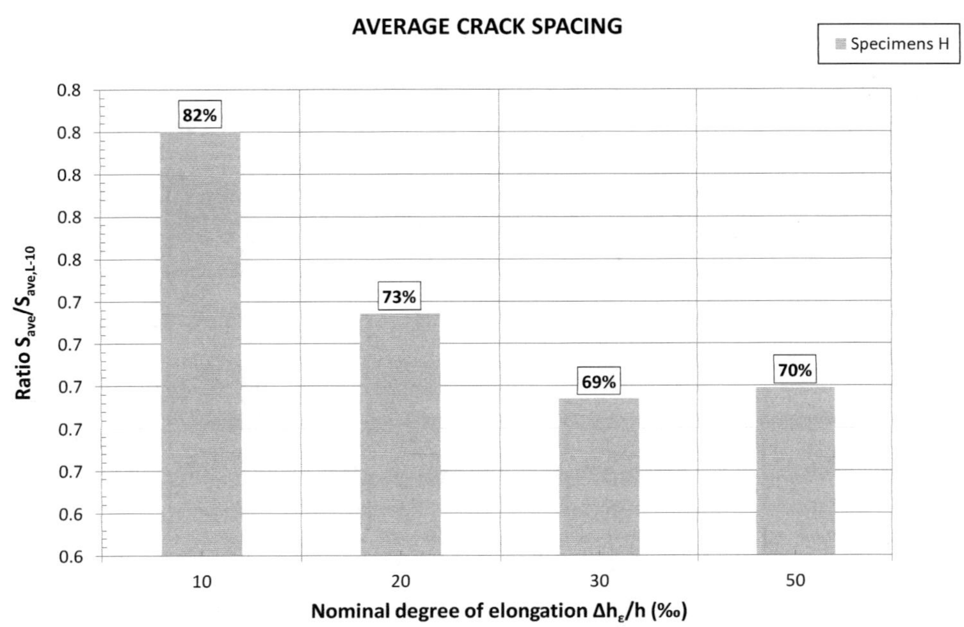


**Figure 13.** Bar chart of average crack spacing as a percentage of the average crack spacing of the reference specimen.

The experimental findings of the test elements were then analysed and evaluated:

1. The number of cracks appearing increases by one crack for each elongation degree, 10‰, 20‰, 30‰, and remains constant thereafter for the elongation degree of 50‰. For a better understanding of this phenomenon, more experiments concerning different longitudinal reinforcement ratios and arrangements of rebars need to be performed.
2. Comparing the crack width with the degree of elongation applied, it can be seen that the width becomes larger as the elongation degree sustained increases (Table 2, Fig. 5). It is noteworthy that all types of crack width increase with the increment of the tensile degree – meaning the minimum, maximum and average crack width.
3. It seems that there is a general trend for decreased crack spacing compared to the spacing of cracks for the initial degree of elongation of 10‰ (Fig. 6). The explanation for this may be that the number of cracks formed increases for degrees higher than the initial 10‰, which eventually translates to smaller spacings between a larger number of cracks.
4. The spacings remain almost constant between cracks, independent of the degree of elongation.
5. The damage state of specimens indicates that cracks appear at or near to the tie positions (Fig. 4). Thus, the presence of steel ties helps and promotes the disorganization of concrete around them.

## 4. Conclusions

This paper looks at four specimens to investigate cracking formation and behaviour in terms of the number of cracks, their width and spacing. The following conclusions are drawn:

1. The degree of tensile deformation holds a significant part in terms of the formation of cracks and their characteristics, e.g. the number of cracks formed, the width and spacing of cracks – whether considering the minimum, maximum or average value.
2. Higher degrees of elongation result in cracks with larger widths. Thus, the design of reinforced concrete structural components should take into account the degree of elongation because, as it is well known, large crack widths can lead to oxidization and deterioration of rebars and eventually affect structural safety.
3. The spacing between cracks seems to be affected when a higher tensile degree leads to a higher number of cracks. Otherwise, spacing characteristics remain more or less unaffected too.
4. The question arises of whether the longitudinal ratio or whether the arrangement of rebars plays an essential role, too. Further research is needed on the subject using test specimens with different longitudinal reinforcement ratios and arrangements. This will help to check the impact that the mechanical factor of reinforcement ratio has.

## References

[1] H T Hu and W C Schnobrich 1990 "Nonlinear analysis of cracked reinforced concrete" *ACI Structural Journal* **87** 199–207.

[2] Z Shi, M Suzuki and M Nakano 2003 "Numerical Analysis of Multiple Discrete Cracks in Concrete Dams Using Extended Fictitious Crack Model" *Journal of Structural Engineering* **129** pp 324–336.

[3] C Ouyang, E Wollrab, S M Kulkarni and S P Shah 2002 "Prediction of Cracking Response of Reinforced Concrete Tensile Members" *Journal of Structural Engineering* **123** pp 70–78.

[4] K A Vu and M G Stewart 2005 "Predicting the Likelihood and Extent of Reinforced Concrete Corrosion-Induced Cracking" *Journal of Structural Engineering* **131** pp 1681–1689.

[5] M Lorrain, O Maurel and M Seffo 1998 "Cracking behavior of reinforced high-strength concrete tension ties" *ACI Structural Journal* **95** pp 626–635.

[6] S L Lee, M A Mansur, K H Tan and K Kasiraju 1987 "Cracking Behavior of Concrete Tension Members Reinforced With Welded Wire Fabric," **84** pp 481–491.

[7] H Marzouk, M Hossin and A Hussein 2010 "Crack width estimation for concrete plates," *ACI Structural Journal* **107** pp 282–290.

[8] C H Lu, W L Jin and J H Mao 2011 "Experimental Investigation of Corrosion-Induced Cover Cracking in Reinforced Concrete Structures" *Advanced Materials Research* **197–198** pp 1690–1693.
[9] S K Padmarajaiah and A Ramaswamy 2001 "Crack-width prediction for high-strength concrete fully and partially prestressed beam specimens containing steel fibers" *ACI Structural Journal* **98** pp 852–861.
[10] R Gaetano and R Filippo 1992 "Cracking Response of RC Members Subjected to Uniaxial Tension" *Journal of Structural Engineering* **118** pp 1172–1190.
[11] H Gesund and R J Frosh 2000 "Another look at cracking and crack control in reinforced concrete Paper by Robert J Frosh" *ACI Structural Journal* **97** p 355.
[12] B H C Chan, Y K Cheung and Y P Huang 1993 "Crack analysis of reinforced concrete tension members" **118** pp 2118–2132.
[13] H Mirzabozorg and M Ghaemian 2005 "Non-linear behavior of mass concrete in three-dimensional problems using a smeared crack approach" *Earthquake Engineering and Structural Dynamics* **34** pp 247–269.
[14] M Byfield 2018 "Control of cracking in reinforced concrete" *Structural Design from First Principles* **105** pp 271–288.
[15] B P Hughes and C V Cifuentes 1988 "Comparison of Early-Age Crack Width Formulas for Reinforced Concrete" *ACI Structural Journal* **85** pp 158–166.
[16] S Yang and J Chen 1988 "Bond Slip and Crack Width Calculations of Tension Members," *ACI Structural Journal* **85** pp 414–422.
[17] H G Sohn, Y M Lim, K H Yun and G H Kim 2005 "Monitoring crack changes in concrete structures" *Computer-Aided Civil and Infrastructure Engineering* **20** pp 52–61.
[18] M M Smadi and F O Slate 1989 "Microcracking of high and normal strength concretes under short- and long-term loadings" *ACI Materials Journal* **86** pp 117–127.
[19] Y H Lee and K Willam 1997 "Mechanical properties of concrete in uniaxial compression," *ACI Materials Journal* **94** pp 457–471.
[20] M Aschheim and J Browning 2008 "Influence of Cracking on Equivalent SDOF Estimates of RC Frame Drift" *Journal of Structural Engineering* **134** pp 511–517.
[21] G Chen and G Baker 2003 "Influence of Bond Slip on Crack Spacing in Numerical Modeling of Reinforced Concrete" *Journal of Structural Engineering* **129** pp 1514–1521.
[22] N Ariyawardena, A Ghali and M Elbadry 1997 "Experimental study on thermal cracking in reinforced concrete members" *ACI Structural Journal* **94** pp 432–441.
[23] Z P Bažant and B H Oh 2008 "Spacing of Cracks in Reinforced Concrete" *Journal of Structural Engineering* **109** pp 2066–2085.
[24] B Bengt and L Leroy 1965 "Effects of Arrangement of Reinforcement on Crack Width and Spacing of Reinforced Concrete Members" *ACI Journal Proceedings* **62** pp 1395–1420.
[25] Y Goto 1971 "Internal Cracks Formed in Concrete Around Deformed Tension Bars" *ACI Journal Proceedings* **68** pp 244–251.
[26] R R H Zhu, W Wanichakorn, T T C Hsu and J Vogel 2003 "Crack width prediction using compatibility-aided strut-and-tie model" *ACI Structural Journal* **100** pp 413–421.
[27] Wood S L, Wight J K and J P Moehle 1987 "The 1985 Chile Earthquake, Observations on Earthquake Resistant Construction in Viña del Mar" *Civil Engineering Studies* p 176.
[28] M Fintel 2014 "Performance of Buildings With Shear Walls in Earthquakes of the Last Thirty Years" *PCI Journal* **40** pp 62–80.
[29] G G Penelis and A J Kappos 1996 *Earthquake-resistant Concrete Structures* London, UK: E & F N SPON (Chapman & Hall).
[30] G Penelis, K Stylianidis, A Kappos and C Ignatakis 1995 *Reinforced Concrete Structures* Thessaloniki, Greece: AUTh Press.
[31] J W Wallace 2012 "Behavior, design, and modeling of structural walls and coupling beams - Lessons from recent laboratory tests and earthquakes" *International Journal of Concrete*

*Structures and Materials* **6** pp 3–18.
[32] Y H Chai and S K Kunnath 2005 "Minimum thickness for ductile RC structural walls" *Engineering Structures* **27** pp 1052–1063.
[33] Standards New Zealand 2006 "NZS 3101:2006, Concrete structures standard: Part 1 – The design of concrete structures" Wellington, New Zealand.
[34] Canadian Standards Association 2007 "CAN/CSA-A233-04, Design of Concrete Structures (Update No 2 - July 2007)," Mississauga, Ontario, Canada.
[35] European Committee for Standardization 2004 "EN 1998-1:2004, Eurocode 8: Design of structures for earthquake resistance - Part 1: General rules, seismic actions and rules for buildings" Brussels, Belgium.
[36] Ministry of Environment Planning and Public Works 2000 "Greek Code for the Design and Construction of Concrete Works" Athens, Greece. (In Greek)
[37] International Conference of Building Officials 1997 "Uniform Building Code – Volume 2: Structural Engineering Design Provisions" Whittier, California, USA.
[38] T A Chrysanidis 2020 "Evaluation of Out-of-Plane Response of R/C Structural Wall Boundary Edges Detailed with Maximum Code-Prescribed Longitudinal Reinforcement Ratio" *International Journal of Concrete Structures and Materials* **14.**
[39] T Chrysanidis 2019 "Influence of elongation degree on transverse buckling of confined boundary regions of R/C seismic walls" *Construction and Building Materials* **211** pp 703–720.
[40] T Chrysanidis and I Tegos 2020 "Axial and transverse strengthening of R/C circular columns: Conventional and new type of steel and hybrid jackets using high-strength mortar" *Journal of Building Engineering* **30**.

IOP Conf. Series: Materials Science and Engineering **897** (2020) 012005 doi:10.1088/1757-899X/897/1/012005

# Elastoplastic Behavior of Wide Flange Beam - to-Wide Flange Column Welded Connections

**M D Pangandaman[1,2*], C S Taclendo[3,4] and K B Cantila[3]**

[1] College of Engineering Graduate, MSU–Iligan Institute of Technology, Tibanga, Iligan City, Lanao del Norte, Philippines, 9200, Philippines
[2] Mindanao State University, Marawi City, 9700, Philippines
[3] Civil Engineering Department, MSU–Iligan Institute of Technology, Tibanga, Iligan City, Lanao del Norte, 9200, Philippines
[4] Civil Engineering Department, MSU–General Santos, General Santos City, Cotabato, 9500, Philippines
*corresponding author's email: merhannap@gmail.com

**Abstract.** A beam-to-column connection is considered to be one of the most critical sections of various types of load-carrying structures. Many parameters such as the geometrical and material properties of both beams and columns influenced its performance. The purpose of this study is to derive analytical models for the connection's yield and ultimate strength and displacements using the depth of wide flange (WF) beam and WF-column, and the length of WF-beam as design parameters. Experimental tests involved testing of three specimens until reaching their maximum strength. The behavior of the connection observed in the experiment were employed in the finite element analyses (FEA), where the thirty numerical models generated with the parametric script in SolidWorks, provided the design information for the mechanical properties. Lastly, the derivation of analytical formulae used to predict the mechanical properties of wide flange beam-to-wide flange column connections employed regression analysis. In conclusion, (1) There is a good agreement between FEA and test results, (2) The dimensional parameters $L_b/D_b$ and $D_c/L_b$ ratios have been found to significantly characterize the elastic-plastic properties of fully welded beam-to-column connections with stiffeners, and (3) The mechanical properties of the connections can be well predicted by the derived analytical formulae.

## 1. Introduction

The beam-to-column connections assume a vital part in the overall performance of a structure. Consequently, inadequate connections can fail the structure regardless of how adequately designed the structural members are. Parameters such as geometrical and mechanical properties of the material used and detailed arrangement of the column influence the performance of the connection. Thus, changing the geometrical properties of connected elements can modify the status of their stresses, strains, and the structure failure mode [1]. Recent investigations on the elastic-plastic behavior of beam-to-column connection focused on the parametric investigation of a hollow structural column to WF-beam connections and their impact within the structural engineering profession [2-5]. In addition, existing studies that derive formulae that can estimate beam-to-column connections' strength and displacements are relevant only to a wide flange beam-to-hollow structural steel (HSS) column connections. The present study derived parametric formulae that can predict WF beam-to-WF column

IOP Conf. Series: Materials Science and Engineering **897** (2020) 012005 doi:10.1088/1757-899X/897/1/012005

connection's behavior in terms of local strengths and displacements (i.e., yield load, ultimate load, yield deformation, ultimate deformation). WF beam-to-WF column welded connection is a popular type of connection in the Philippines.

Kamba et al. and Hassan et al. [3-4] obtained analytic formulae through regression analysis that effectively predicted the behavior of connections in a numerical study on circular hollow sections (CHS) connections. Simulation of numerical models under three loading conditions, a compression load, tension load and moment, used the parameters diameter to thickness ratio $(D/t)$ and the flange width to diameter ratio $(B_f/D)$ [3], length of gusset plate, and connection configuration [4]. In a nutshell, the behavior of CHS connections is dependent upon these parameters.

The aforementioned discussion highlights the need to develop analytical models that can predict well the elastic-plastic behavior of beam-to-column connections, particularly beams and columns made of wide flange sections. Lack of understanding, poor design of connection and ignoring the actual response of connection in the structural analysis may lead to inadequate structural design and consequently lead to disaster [6]. In response to this problem, the study proposes to determine the elastic-plastic behavior of W-beam-to-W-column moment connections by obtaining analytical formulae. The findings apply to conventional frame analysis to predict the realistic connection performance of steel frames. Thus, a new theory on building design may be arrived at.

## 2. Materials and Methods

### *2.1. Experimental Program*

*2.1.1. Materials.* To conduct a full-scale experimental test, a test rig was considered to accommodate a smaller specimen, 2.5- meter-height columns (WF 8x24) and 1.6-meter-span cantilever beams (WF 4x13, WF 6x15 and WF 8x24).

Experimental materials used were A36 wide flange sections and A36 steel plates. The beam-to-column connections were designed in accordance with the American Institute of Steel Construction [6-8], and National Structural Code of the Philippines [9], while the horizontal stiffeners followed the American Welding Society [10-11] specification. In the WF specimens, the beam flanges were connected to the column flange, and the sides of the beam web to the column flange by fillet welds. Details of the connections and their configurations are shown in Table 1and Figure 1, respectively.

**Table 1.** Details of Specimens for Experimental Tests.

| Specimen | Column | Beam | Length of the beam (mm) | Fillet weld: Size (mm) for A36 steel | Fillet weld: Length in each side of the web (in) | Fillet weld: Length of flange weld (in.) |
|---|---|---|---|---|---|---|
| m14-0.15 | WF 8x24 | WF4x13 | 1600 | 6 | 6.33 | 4.06 |
| m9-0.15 | | WF 6x15 | 1600 | 6 | 4.97 | 5.99 |
| m7-0.15 | | WF 8x24 | 1600 | 6 | 8.85 | 6.495 |

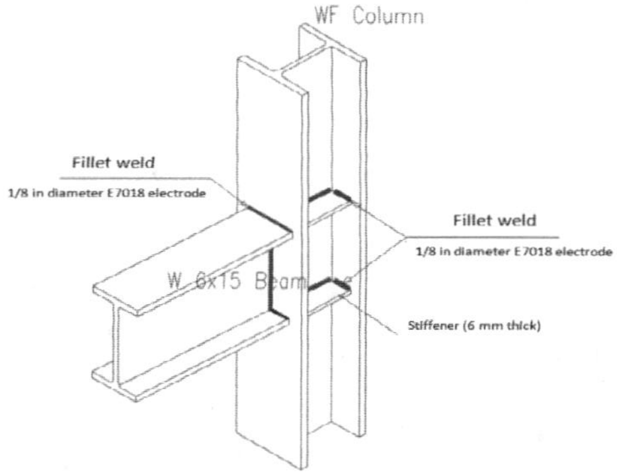


**Figure 1.** Configuration of the Specimens.

To avoid confusion, the names of the specimens were matched to their numerical designation.

*2.1.2. Test Procedure.* Data acquisition of the load applied on the beam through the load cell involved the use of LabView software and National Instruments (NI) data acquisition devices. As shown in Figure 2, a compression load cell was connected to the NI 9219 universal analog input module, then plugged into the NI CompactDAQ system and connected back into a laptop. A steel ruler was positioned at the loading frame to measure the vertical deflection of the specimens at increasing load.

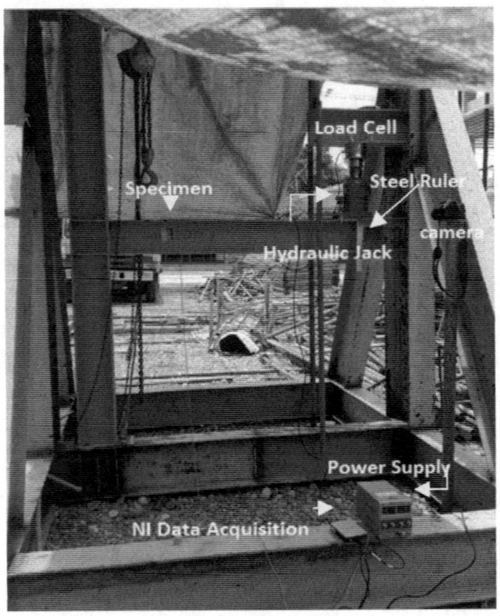


**Figure 2.** Test Setup.

A recording of the initial zero-load reading began the test. Loads were applied to the cantilever beam at 1.4 m from the face of the column using a hydraulic jack. The loading of the specimen was increased until it reached the ultimate strength.

*2.2 Numerical Modeling*

*2.2.1. Finite Element Model.* Finite element models of beam-to-column connections under monotonic loads were analyzed using SolidWorks 2016 software.

A set of numerical models were developed according to the tests' set-up, respecting the same geometry, boundary conditions and the applied loading in the experimental tests.

A displacement control nonlinear static analysis was performed, employing a Von-Mises plastic model type to accurately simulate the plastic behavior of the connections. In FE simulations, the values of material properties of A36 steel in a nonlinear simulation were the default values provided SolidWorks software except for the Tangent Modulus that was inputted through the Material dialog box.

## 3. Results and Discussion

*3.1. Experimental Test Results*

Force and beam's deformation at yielding and ultimate points were determined through experimental investigations. Table 2 shows the elastic-plastic properties of the three specimens considered. Bending stresses $\sigma$ were taken from theoretical computations, employing the formula:

$$\sigma = PL_b/S_x \qquad [1]$$

where $\sigma$= bending strength of the beam, $S_x$ = Section modulus of the beam along the axis of the beam, and
$L_b$ = distance from the face of the column to the point along the beam where the load is applied.

**Table 2.** Summary of Test Results (Actual).

| Specimen | $PL$ (MPa) | $\boldsymbol{P_y}$ (kN) | $\boldsymbol{P_u}$ (kN) | $\boldsymbol{\delta_{PL}}$ (mm) | $\boldsymbol{\delta_y}$ (mm) | $\boldsymbol{\delta_u}$ (mm) | $\boldsymbol{\sigma_y}$ (MPa) | $\boldsymbol{\sigma_u}$ (MPa) |
|---|---|---|---|---|---|---|---|---|
| m14-0.15 | 18.65 | 19.35 | 25.61 | 39.69 | 44.45 | 252.12 | 304.26 | 402.77 |
| m9-0.15 | 27.54 | 29.76 | 49.32 | 38.10 | 41.28 | 139.70 | 344.60 | 570.99 |
| m7-0.15 | 29.96 | 39.76 | 74.45 | 31.75 | 52.39 | 257.81 | 162.55 | 304.58 |

The design weld size for all connections was the maximum weld theoretically computed. This is the prime reason the connections did not show visible cracks. At the early stage of loading, m14-0.15 already manifested small lateral buckling to its right. There was also a noticeable deformation in the panel zones, especially in m7-0.15 and m9-0.15 since larger loads were applied to the two connections to deform them. Added column stiffeners added strength to the columns as well.

*3.1.1. Summary of the Experimental Tests.* The main findings of the experimental investigations were the $P_u$/ $P_y$ (= 1.32, 1.66 and 1.87) and the Proportional Limit (= $0.63P_y$, $0.49P_y$and $0.75P_y$) ratios of the specimens m14-0.15, m9-0.15, and m7-0.15, respectively. These ratios will form the points A, B, and C of the trilinear curves of the numerical models, following the proposed trilinear model of Kamba and Taclendo [3] and Hassan et al. [4]. Taking the average $P_u$/ $P_y$ of the three specimens gave 1.62. $P_u$, the load required to reach the maximum load a beam can carry. The Proportional limit is the point where the connection behaves elastically. This is the 1st point of the load-deformation curve and is located at an average of $0.62P_y$.

*3.1.2. Comparison between Experimental and FEA Results.* Figures 3 (a) to (c) show the von Mises stress distribution on the assemblies of m14-0.15, m9-0.15 and m7-0.15 simulated in SolidWorks. The variation of colors represents the intensity of the stress developed at all sections of the assembly and the magnitude of which is shown in the color bar at the right.

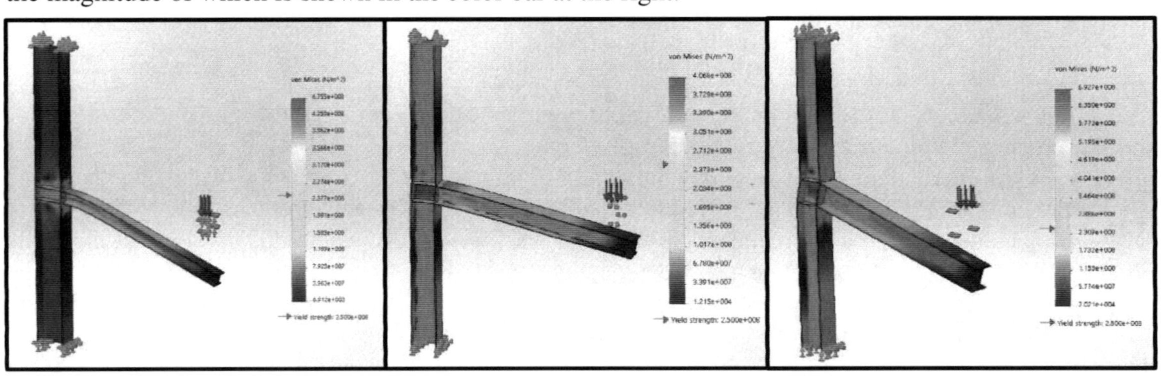

(a) (b) (c)

**Figure 3.** FEA Von-Mises stress distribution for (a) m14-0.15, (b) m9-0.15 and (c) m7-0.15.

Graphs in Figure 4 summarize the comparison of load-deformation (P-Δ) of the experimental to the FEA results. *PL*, $P_y$ and $P_u$ are represented in the curves as circles and triangles. The circles indicate the results in the experimental tests while the triangles indicate the FEA results. It can be observed that ultimate loads from experimental investigation and finite element simulations have closer values than that of yield loads. Further, it can be seen in the curves that FEA well-estimated m14-0.15, underestimated m9-0.15, and overestimated m7-0.15 a little to the test results. However, good correspondences have been found between the experimental stresses and FEA stresses, hence, it can be safely said that there is a good agreement between the experimental and numerical results, particularly in the plastic domain.

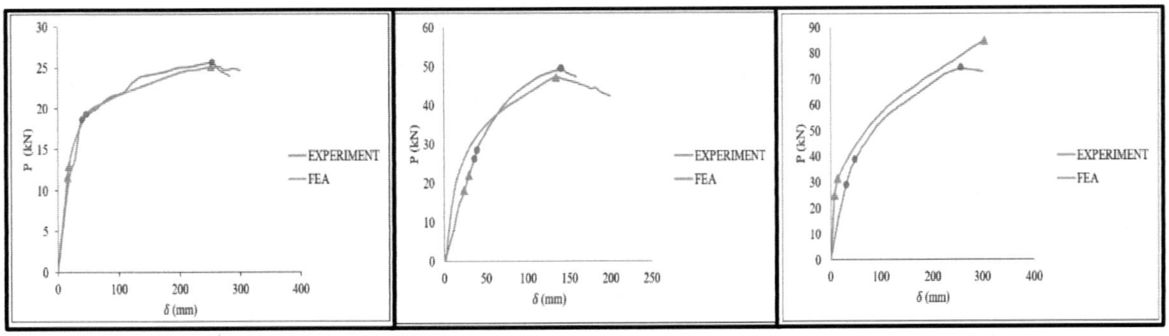


(a) (b) (c)

**Figure 4.** Comparison of Experimental and FEA results (a) m14-0.15, (b) m9-0.15 and (c) m7-0.15.

### *3.2 Finite Element Analysis Results*

Tables 3 shows the mechanical properties of beam-to-column connections obtained from finite element simulations of 27 models. In the analyses, length of the beam to its depth and the depth of the column to the beam's length ($L_b/D_b$ , $D_c/L_b$) were chosen as design parameters where $L_b$ the length of the beam is, $D_b$ is the depth of the Wide Flange beam, and $D_c$ is the depth of the Wide Flange column. The designation of specimens was taken from the loading configuration- monotonic loading. Specimens under monotonic loadings were represented by the prefix, "m". The first numerical designation, i.e., 14, 9, 7, 5.5, 4.6, and 4, represented the ratio of the beam's length over its depth. The ratio of the column's depth over the beam's length of the specimen was indicated by numbers 0.15, 0.18, 0.22, 0.25 and 0.29. The results of the experimental test conducted on three specimens are also shown in the table marked * at the specimen's column.

**Table 3.** FEM Results of specimens.

| SPECIMEN | COLUMN | BEAM | PL (kN) | $\delta_{PL}$ (mm) | Py (kN) | $\delta_y$ (mm) | Pu (kN) | $\delta_u$ (mm) |
|---|---|---|---|---|---|---|---|---|
| m14-0.15* | WF 8x24 | WF 4x13 | 18.65 | 39.69 | 19.35 | 44.45 | 25.61 | 252.12 |
| m9-0.15* | | WF 6x15 | 27.54 | 38.10 | 29.76 | 41.28 | 48.22 | 139.7 |
| m7-0.15* | | WF 8x24 | 29.96 | 31.75 | 39.76 | 52.39 | 74.45 | 258.81 |
| m5.5-0.15 | | WF 10x22 | 34.38 | 9.97 | 55.45 | 29.33 | 89.83 | 92.08 |
| m4.6-0.15 | | WF 12x26 | 43.92 | 8.09 | 70.84 | 12.4 | 114.76 | 24.8 |
| m4-0.15 | | WF 14x22 | 48.16 | 8.36 | 77.68 | 13.93 | 125.84 | 23.35 |
| m14-0.18 | WF 10x26 | WF 4x13 | 12.06 | 15.74 | 19.46 | 37.04 | 31.52 | 145.37 |
| m9-0.18 | | WF 6x15 | 22.83 | 16.84 | 36.82 | 37.83 | 59.65 | 201.54 |
| m7-0.18 | | WF 8x24 | 30.26 | 8.71 | 48.8 | 20.52 | 79.06 | 63.12 |
| m5.5-0.18 | | WF 10x22 | 46.71 | 11.33 | 75.34 | 28.33 | 122.05 | 118.98 |
| m4-0.18 | | WF 14x22 | 62.77 | 9.03 | 101.24 | 14.64 | 164.01 | 27.21 |
| m14-0.22 | WF 12x30 | WF 4x13 | 12.36 | 16.92 | 19.93 | 37.04 | 32.29 | 279.38 |
| m9-0.22 | | WF 6x15 | 22.98 | 14.09 | 37.06 | 37.77 | 60.04 | 212.23 |
| m7-0.22 | | WF 8x24 | 38.33 | 11.3 | 61.82 | 32.86 | 100.15 | 133.5 |
| m5.5-0.22 | | WF 10x22 | 51.03 | 11.44 | 82.31 | 28.61 | 133.34 | 104.15 |
| m4.6-0.22 | | WF 12x26 | 67.85 | 9.14 | 109.44 | 16.44 | 177.29 | 30.46 |
| m4-0.22 | | WF 14x22 | 66.99 | 9.71 | 108.05 | 18.03 | 175.04 | 33.29 |
| m14-0.25 | WF 14x30 | WF 4x13 | 14.04 | 19.16 | 22.64 | 57.49 | 36.68 | 320.36 |

| | | | | | | | | |
|---|---|---|---|---|---|---|---|---|
| m9-0.25 | | WF 6x15 | 23.44 | 14.33 | 37.80 | 37.67 | 61.24 | 217.33 |
| m7-0.25 | | WF 8x24 | 57.20 | 16.58 | 92.36 | 32.64 | 149.62 | 63.21 |
| m5.5-0.25 | | WF 10x22 | 58.22 | 10.82 | 86.68 | 28.45 | 140.42 | 110.95 |
| m4.6-0.25 | | WF 12x26 | 80.12 | 10.46 | 129.23 | 18.3 | 209.35 | 33.99 |
| m4-0.25 | | WF 14x22 | 80.64 | 9.31 | 130.07 | 17.22 | 210.71 | 33.16 |
| m14-0.29 | WF 16x31 | WF 4x13 | 14.25 | 19.83 | 22.98 | 59.96 | 37.23 | 324.66 |
| m9-0.29 | | WF 6x15 | 24.80 | 13.85 | 40.00 | 37.69 | 64.80 | 161.73 |
| m7-0.29 | | WF 8x24 | 60.90 | 15.14 | 98.23 | 32.81 | 159.13 | 65.62 |
| m5.5-0.29 | | WF 10x22 | 58.22 | 10.73 | 93.9 | 29.21 | 152.12 | 118.03 |
| m4.6-0.29 | | WF 12x26 | 81.23 | 9.25 | 131.01 | 16.76 | 212.24 | 31.21 |
| m4-0.29 | | WF 14x22 | 86.12 | 9.53 | 138.90 | 17.60 | 225.02 | 41.06 |

*3.3 Analytical Analysis of the Numerical data*

*3.3.1. Relationship between Applied Load P and $L_b/D_b$ and $D_c/L_b$* The larger the $D_c/L_b$ ratio and smaller the $L_b/D_b$ ratio, the higher the strength the specimens have. This is evident in the graphs of Py and $D_c/L_b$ and Pu and $D_c/L_b$ in Figure 5 (a) and (b). Power trendlines in Microsoft Excel were also added to give a clear illustration of the effect on the mechanical properties of the connections by the predictors. The high values of the Pearson correlation coefficient $r$ significantly indicate a strong correlation of variables. In summary, the proportional limit, yield and ultimate strengths of the specimens displayed similar behavioral pattern and close values of $r$.

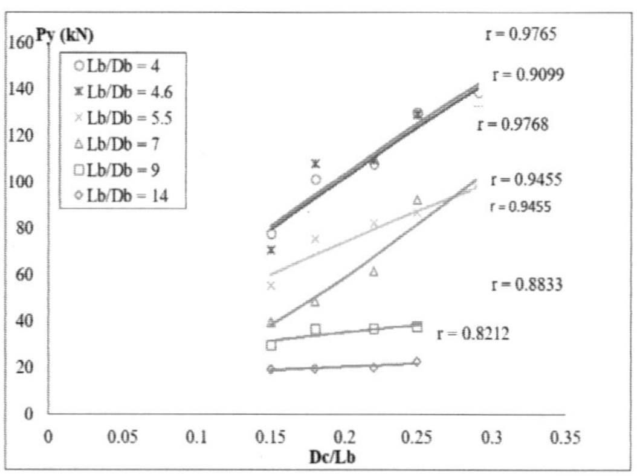


(a)

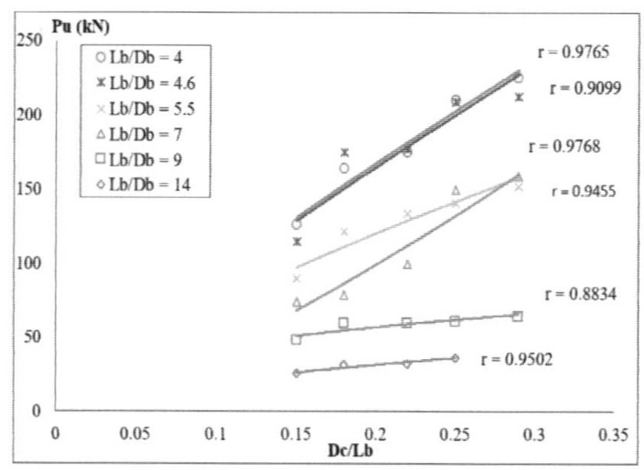


(b)

**Figure 5.** Relationship between Load Applied and $D_c/L_b$ (a) Py vs. $D_c/L_b$ and (b) Pu vs. $D_c/L_b$.

*3.3.2. Relationship between Deformation $\delta$ and $L_b/D_b$, $D_c/L_b$* Generally, the larger the $L_b/D_b$ and $D_c/L_b$ ratios, the larger the deformation as shown in both figures in Figure 6 a, b. The graphs below exhibit the direct relationship between deformation and the predictors. In the $D_c/L_b$ ratios, 0.18 showed the least deformation in the graph despite 0.15 being the lowest $D_c/L_b$ ratio. However, in a general sense, the deformation of the specimens increases as the $D_c/L_b$ ratio increases.

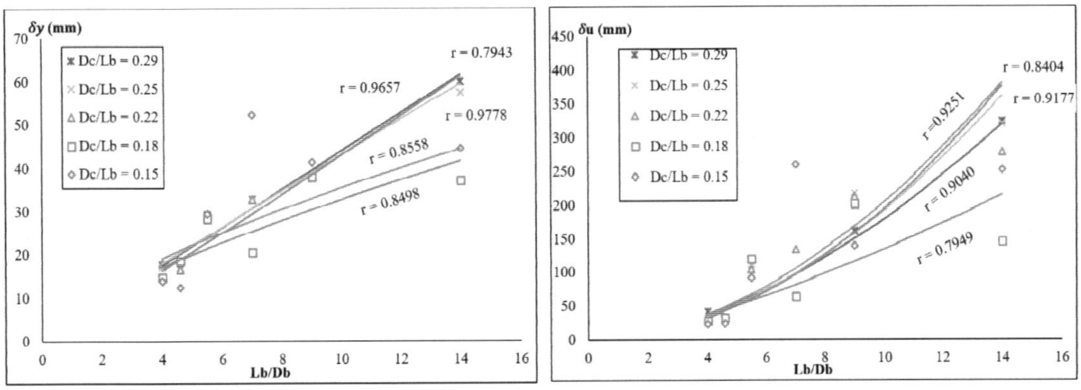


(a) (b)

**Figure 6.** Relationship between Deformation and $L_b/D_b$ (a) $\boldsymbol{\delta_y}$ vs. $L_b/D_b$ and (b) $\boldsymbol{\delta_U}$ vs. $L_b/D_b$.

*3.3.3. Modeling the mechanical properties of beam-to-column connections.* Dimensional analysis, per se, analyzes the relationships between physical quantities. A physical quantity is represented non-dimensionally by a non-dimensional product of all physical quantities related to it. For example, the physical quantities related to the yielding load Py of the connections are the length to the depth of the beam ratio and the depth of the column to the length of the beam ratio as per the high value of correlation coefficients. In the analytical modeling, these physical quantities were taken into account to obtain values closer to the test and FEA results.

Considering the in-depth analysis on the relationship between the mechanical properties of beam-to-column connections to the geometrical and material parameters performed in the previous sections, the yield load parameter is

$$P_y = X_1(L_b/D_b)^a X_2(D_c/L_b)^b X_3 \varepsilon$$

where, $\varepsilon$ is the error. Dependent variables are the mechanical properties of the connection while the independent variables are the geometric parameters $L_b/D_b$ and $D_c/L_b$. A power formula was used because it is the best-fit curve for the FEA and test data. The numerical values for $X_1$, $X_2$, $X_3$, a, and b were determined using the method of least squares with 3 test and 27 FEA results. Performing the said statistical technique for all mechanical properties, the analytical models are the following:

$$P_y = \sigma_y {L_b}^2[0.0047(L_b/D_b)^{-1.458}(D_c/L_b)^{0.582}] \quad \text{(N)} \tag{2}$$
$$P_u = \sigma_y {L_b}^2[0.0067(L_b/D_b)^{-1.42}(D_c/L_b)^{0.59}] \quad \text{(N)} \tag{3}$$
$$\delta_y = L_b[6.5754E^{-6}(L_b/D_b)^{0.7412}(D_c/L_b)^{0.1726}] \quad \text{(mm)} \tag{4}$$
$$\delta_u = L_b[9.1012E^{-6}(L_b/D_b)^{1.2962}(D_c/L_b)^{0.2189}] \quad \text{(mm)} \tag{5}$$

where, $\sigma_y$ is the yield strength of the beam and column section expressed in terms of $N/mm^2$ and $L_b$ is the length of the WF beam in millimeter, $D_b$ is the depth of the WF beam in millimeter and $D_c$ is the depth of the WF column in millimeter.

*3.3.4. Comparison between Experiment, FEA and Predicted results.* The comparison of the load-deformation curves of the test (m14-0.15, m9-0.15 and m7-0.15) and FEA to the predicted values are shown in Figures 7 (a), (b) and (c). The trilinear curve models were plotted using the analytical models derived in the previous section. The predicted values tend to overestimate the experimental and FEA results at a smaller $L_b/D_b$ ratio. It is evident in the graphs that the analytical models yielded good prediction for m14-0.15 ($L_b/D_b = 14$), but these overestimated prediction for specimens m9-0.15 ($L_b/D_b = 9$) and m7-0.15 ($L_b/D_b = 7$).

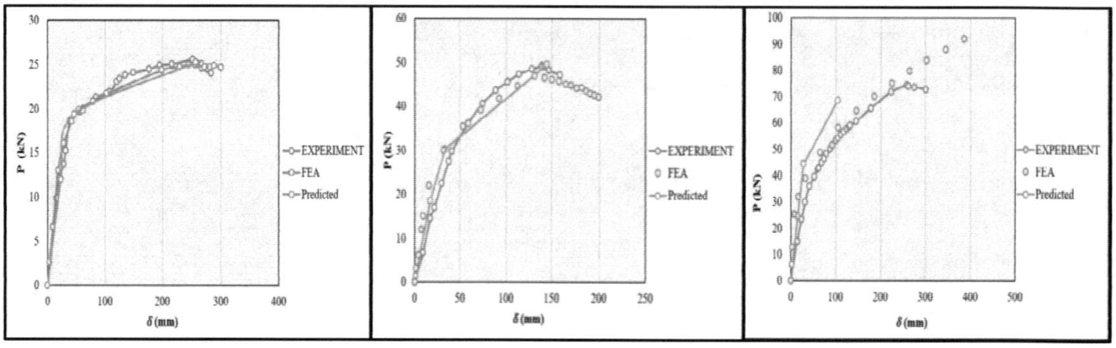


(a) m14-0.15 (b) m9-0.15 (c) m7-0.15

**Figure 7.** Comparison of the Estimated Results to the FEA, and Test Results.

## 4. Conclusions

1) There is a good agreement between FEA and test results, particularly in the plastic domain.
2) The dimensional parameters $L_b/D_b$ and $D_c/L_b$ ratios have significantly characterized the elastic-plastic properties of fully welded beam-to-column connections with stiffeners.
3) The mechanical properties of the connections can be well predicted by the derived analytical formulae obtained by the nonlinear regression analysis.
4) There is a good correlation between the FEA results and the derived analytical formulae.

**Acknowledgments**

This work was supported by the Department of Science and Technology- Engineering Research and Development for Technology.

## References

[1] Warmka G R and Truman K Z 2000 Improving the performance of steel beam-column moment resistant connections *12th World Conf. on Earthquake Eng.* (Upper Hutt: New Zealand Society for Earthquake Engineering)

[2] Murakami Y and Kaneko T 1996 *Elastic-Plastic Behaviour and Design of Beam-to-Column Connections Reinforced by Increased Thickness of Columns* JFE Steel Corporation **35**

[3] Kamba T and Taclendo C 1998 CHS Column Connections without Stiffener *Proc. of the Eight Int. Symp. on Tubular Structures* (Rotterdam: A.A. Balkema) p 567

[4] Hassan M M, Ramadan, H, Abdel-Mooty, M, and Mourad, SA 2015 Experimental and numerical study of one-sided branch plate-to-circular hollow section connections *Steel and Compos Struct* **19**

[5] Songa QY, Heidarpourb A, Zhaob X L, Hanc, LH 2017 Performance of flange-welded/web-bolted steel I-beam to hollow tubular column connections under seismic load *Thin-Walled Struct* **116** pp 250–264

[6] American Institute of Steel Construction (AISC) 2016 *Specification for Structural Steel Buildings* (Chicago, IL: American Institute of Steel Construction)

[7] American Institute of Steel Construction (AISC). *Seismic Provisions for Structural Steel Buildings* ( Chicago, IL: American Institute of Steel Construction)

[8] American Institute of Steel Construction (AISC) *Stiffening of Wide-Flange Columns at Moment Connections: Wind and Seismic Applications* (Chicago, IL: American Institute of Steel Construction; 2003.

[8] Association of Structural Engineers of the Philippines (ASEP) 2015 *National Structural Code of the Philippines (NSCP)* 1

[10] American Welding Society 2002 *Structural Welding Code – Steel* (Miami FL, USA:
[11] AWS 2000 *Structural Welding Code – Seismic Supplement* (Miami FL, USA: American Welding Society)

IOP Conf. Series: Materials Science and Engineering **897** (2020) 012006 doi:10.1088/1757-899X/897/1/012006

# MOFs-derived $Co_3O_4$ loaded hollow $In_2O_3$ nanofibers with greatly enhanced acetone sensing performance

**L Zhu[1,*], X Chang[1,2] and W B Lu[2]**

[1] School of Materials Science and Engineering, China University of Petroleum, Qingdao 266555, Shandong, People's Republic of China

[2] State Key Laboratory of Heavy Oil Processing, China University of Petroleum, Qingdao 266580, Shandong, P.R. China

*Corresponding author's e-mail: zhulei@upc.edu.cn

**Abstract.** In this paper, metal-organic frameworks (MOFs)-derived $Co_3O_4$ were easily functionalized on $In_2O_3$ nanofibers (NFs) by electrospinning followed a calcination at 600 °C in air. The influences of doping content, morphology, and operating temperature on the acetone sensing performances of $In_2O_3$ based NFs were comprehensively studied. Our results showed that the 20 wt% $Co_3O_4$-$In_2O_3$ NFs exhibited outstanding acetone sensing properties at 300 °C. Exactly, it showed a high response to acetone (S~43.67 @ 100 ppm acetone), which was ~22 times larger than that of pristine $In_2O_3$ NFs. On the same time, it exhibited a short response time (~13 s). The excellent acetone sensing properties of MOFs-derived $Co_3O_4$-$In_2O_3$ NFs can be explained by the formation of p-n junction between $Co_3O_4$ and $In_2O_3$. Our work offers a facile and cost-effective approach to load MOFs-derived $Co_3O_4$ on other metal oxides based NFs as a promising gas sensing material and shows the prospects of MOFs-derived metal oxides loaded in metal oxide NFs used for gas sensors.

## 1. Introduction

As a good solvent, acetone is widely used in the industry. However, acetone is extremely volatile and flammable when the concentration is over 450 $mg/m^3$ [1]. Therefore, it is necessary to explore an accurate acetone sensor. In recent years, metal oxide semiconductors (MOS), including $Co_3O_4$, $In_2O_3$, $WO_3$ have a lot of advantages, such as high chemical stability, easy mass-production and suitability for much target gases, and thus have been largely exploited and used as acetone sensors. Among them, $In_2O_3$ has been widely studied for the detection of both oxidizing gases and reducing gases [2]. However, the pristine $In_2O_3$ shows low gas response and poor selectivity. To overcome this obstacle, two main methods are adopted to enhance the gas sensing performance of $In_2O_3$ based sensors.

On the one hand, it is well known that the gas sensors are operated by the surface reaction of target gas on the MOS materials. Thus, the microstructural parameters of the MOS based sensing films, including specific surface area, porosity and pore size have great influences on the sensing performance of MOS based sensors. Electrospinning has been a common method to fabricate the MOS nanofibers (NFs), which possess high gas sensing performance due to unique one dimensional structure. Such MOS NFs are made of metal oxide nanoparticles and possess high gas accessibility due to its open structures, high porosity and high aspect ratio [3].

On the other hand, various suitable catalysts and dopants have been incorporated with MOS to enhance their sensing performance, such as another metal oxides [4] and noble metals [5]. Among them, MOS are more widely used because of the low-cost and the formation of heterojunction. Recently, Metal-organic frameworks (MOFs) derived hollow and porous metal oxides have been used for gas sensing, due to their large number of reaction sites and convenient gas accessibility [6, 7].

Herein, we synthesized $Co_3O_4$-loaded $In_2O_3$ NFs (using ZIF-67 as templates) with porous hollow 1D nanostructure and $Co_3O_4/In_2O_3$ junctions, and investigated its acetone sensing properties. The results demonstrated that compared with pristine $In_2O_3$ NFs, the $Co_3O_4$-$In_2O_3$ NFs exhibited greatly enhanced acetone sensitivity. Furthermore, we discussed the mechanism for the improved acetone sensing performance.

## 2 Experimental Details

### 2.1 *Synthesis of ZIF-67*

Typically, 0.45 g $Co(NO_3)_2{\cdot}6H_2O$ (AR, 99%, Aladdin) was dispersed in 3 mL water. In the meantime, 5.5 g 2-Methylimidazole (2-MI, 99%, Sigma-Aldrich) was dispersed in 20 mL water and stirring for 20 min at 25 °C, the $Co(NO_3)_2{\cdot}6H_2O$ solution was added into the 2-MI solution and aged for 6 h. Afterwards, the purple precipitates were collected via centrifugation, and dried for 1 days at 80 °C in air.

### 2.2 *Preparation of $Co_3O_4$-$In_2O_3$ NFs*

The $Co_3O_4$-$In_2O_3$ NFs were prepared by electrospinning followed by calcination. $In(NO_3)_3{\cdot}xH_2O$ (99.9%, Aladdin) and ZIF-67 (total weight of 0.2 g) were dispersed in a mixed solution of 2.21 g ethanol and 2.21 g DMF (Sinopharm Chemical Reagent Co. Ltd), and stirring for 1 h at 25 °C. The weight ratios of ZIF-67 were 0, 10 wt%, 20 wt% and 30 wt%, respectively. Afterwards, 0.4 g PVP was injected into above solution and then stirring for 10 h. Subsequently, using a 20-gauge stainless steel needle, the prepared solution was injected into a 5 mL plastic syringe and electro-spun for 4 h. The flow rate was set as 1 mL/h and the applied voltage from needle to the collection roller was set as 18 kV. After that, the obtained NFs were dried for half of the day at 60 °C in air. In order to remove the PVP, convert the $In(NO_3)_3$ to the $In_2O_3$ and convert the ZIF-67 to the $Co_3O_4$, the dried NFs were heated at 600 °C for 3 h in air.

### 2.3 *Samples fabrication and acetone sensing measurement*

The 20 mg above-obtained samples were dispersed into 10 mL deionized water to prepare solution. And then we dropped 10 μL solution on the Pt interdigitated electrodes. Afterwards, the $In_2O_3$ NFs based sensing layers on the substrate were dried at 120 °C for 20 min. Finally, all samples were aged in the air for 2 h at 400 °C, which could be used to improve the stability and reproducibility of obtained samples. The process steps in the fabrication of our $In_2O_3$ based sensors and the gas-sensing

IOP Conf. Series: Materials Science and Engineering **897** (2020) 012006 doi:10.1088/1757-899X/897/1/012006

test platform are illustrated in Figure 1.

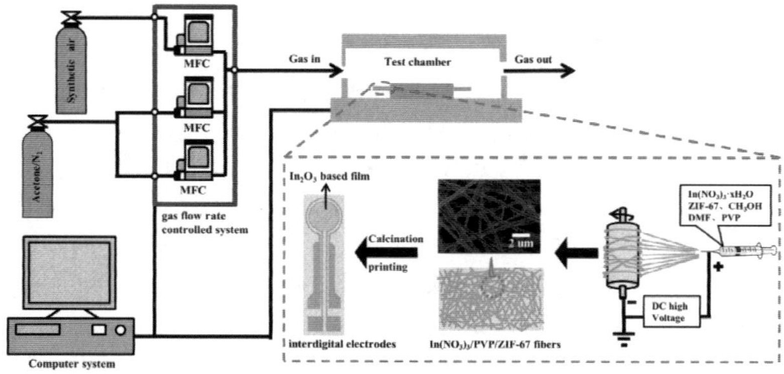


**Figure 1.** The schematic diagram of the gas-sensing test system and fabrication of $In_2O_3$ based sensors.

The acetone sensing characteristics were carried out by a commercial gas detection apparatus. The detection apparatus consists of electrical signal testing system, resistive-type heater system and gas flow controlled system. Using four computer-driven digital mass flow controllers, the gas concentration was controlled by changing the mixing ratio of standard acetone (acetone in $N_2$, Qingdao Ai Ruo Gas Co. Ltd) and synthetic air. The total constant gas flow was set as 1 L/min. The response was calculated by $S = R_a/R_g$, where $R_g$ represents the resistance of the sensors when exposed to the target gas and $R_a$ represents the resistance of the sensors in the background atmosphere.

2.4 *Characterization of the samples*

To investigate the crystallographic structure of samples, X-ray diffraction (XRD) was conducted through Cu-K$\alpha$ radiation ($\lambda$ = 1.5406 Å) in a wide angle range (2θ = 10–90°). The samples' surface morphology and microstructures were characterized by transmission electron microscope (TEM; JEM-2100) and scanning electron microscopy (SEM; Hitachi S-4800). The chemical composition and states were studied through X-ray photoelectron spectroscopy (XPS, ULVAC-PHI5000). The specific surface areas of the NFs were investigated using the Brunauer-Wmmett-Teller (BET) method (TriStar 3020 ).

## 3 Results and discussion

3.1 *Structure characterization of the $Co_3O_4$-$In_2O_3$ NFs*

Firstly, we investigated the phase composition and crystal structure of the samples by XRD measurement. Figure 2 illustrates the XRD patterns of pure $In_2O_3$ NFs and $Co_3O_4$ doped $In_2O_3$ NFs. As shown in Figure 2 (a), the peaks of the pure $In_2O_3$ NFs could be well indexed to crystal structure of $In_2O_3$ (JCPDS: 06-0416) [2]. Besides, it can be confirmed that the $In(NO_3)_3$ are totally converted to $In_2O_3$ after calcination, due to that no other diffraction peaks are observed. Figure 2 (b-d) shows the XRD pattern of 10, 20, 30 wt% $Co_3O_4$-$In_2O_3$. It is shown that the characteristic peaks of both $In_2O_3$ and $Co_3O_4$ can be observed. Besides, these two peaks matches well with the cubic $In_2O_3$ and the cubic $Co_3O_4$ (JCPDS: 43-1003) [8]. The peaks around 2θ = 36.84°, 59.35° and 65.23° are indexed to the (311), (511) and (440) planes of crystal $Co_3O_4$, demonstrating that the ZIF-67 are totally converted into

$Co_3O_4$. The peaks of $Co_3O_4$ are weak in $Co_3O_4$-$In_2O_3$ NFs, probably resulting from that the quantity of ZIF-67 is too little. In addition, it is obvious that with increasing ZIF-67 doping content, the intensity of the diffraction peaks of $Co_3O_4$ increases while the intensity of the diffraction peaks of $In_2O_3$ decreases.

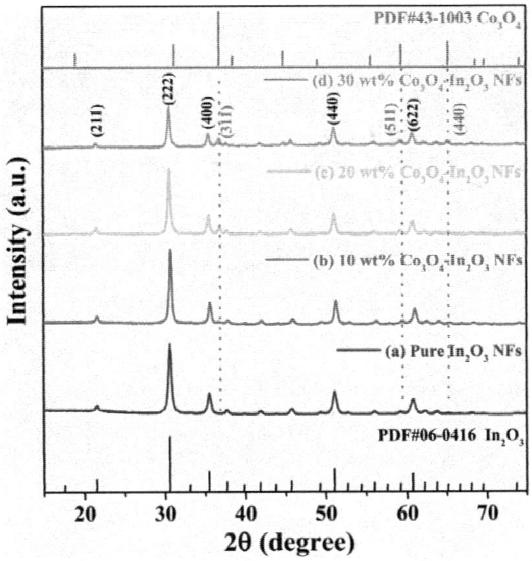


**Figure 2.** XRD patterns of (a) pure $In_2O_3$ NFs, (b) 10 wt% $Co_3O_4$-$In_2O_3$ NFs, (c) 20 wt% $Co_3O_4$-$In_2O_3$ NFs, (d) 30 wt% $Co_3O_4$-$In_2O_3$ NFs.

The morphology and size of the as-prepared ZIF-67 crystals were characterized by SEM. From Figure 3 (a) and (b), it is shown that ZIF-67 crystals are typical polyhedrons and the sizes of them are from tens to hundreds nanometers. The SEM images of pure $In_2O_3$ NFs and 20 wt% $Co_3O_4$-$In_2O_3$ NFs are shown in Figure 3 (c) and (e), respectively. The diameter of all $In_2O_3$ based NFs are about 100~200 nm and they are composed of strings of tiny nanoparticles after heat treatment at 600 °C. It is observed that the surface of pure $In_2O_3$ NFs is smooth and the heating treatment results in a thick-walled tubular structure, which is well correlated with the TEM result in Figure 3 (d). The inset picture in Figure 3 (d) shows the selected area electron diffraction (SAED) patterns of pure $In_2O_3$ NFs. It is shown that the $In_2O_3$ NFs consists of a series of concentric rings, indicating the $In_2O_3$ NFs possess high crystallinity. Compared with the pristine $In_2O_3$ NFs, the 20 wt% $Co_3O_4$-$In_2O_3$ NFs shown in Figure 3 (e) and (f) have larger surface-to-volume ratios and higher porosity because the ultra-thin walled tubular structures and numerous microporous on it. Due to that this 1D hollow NFs structures possess highly porous morphology, the acetone molecules can easily get inside and react with sensing materials. The SAED pattern of 20 wt% $Co_3O_4$-$In_2O_3$ NFs is similar with that of the pure $In_2O_3$ NFs and reveals polycrystalline nature of $In_2O_3$ and $Co_3O_4$ nanoparticles. The HRTEM image of $In_2O_3$ NFs illustrates that the lattice-fringes spacing is 0.292 nm, which is corresponding to the $In_2O_3$ (222) crystallographic planes (see Figure 3 (g)).[9] From Figure 3 (h), it can be observed that both crystalline $In_2O_3$ and $Co_3O_4$ nanoparticles are included in 20 wt% $Co_3O_4$-$In_2O_3$ NFs. Besides, the lattice fringes spacing of 0.292 nm and 0.467 nm are corresponding to the (222) plane of the $In_2O_3$ and (111) plane of the $Co_3O_4$, respectively. Therefore, it is sure that the $In(NO_3)_3$ and ZIF-67 are totally converted to $In_2O_3$ and $Co_3O_4$ after heat treatment, which is consistent with the XRD results.

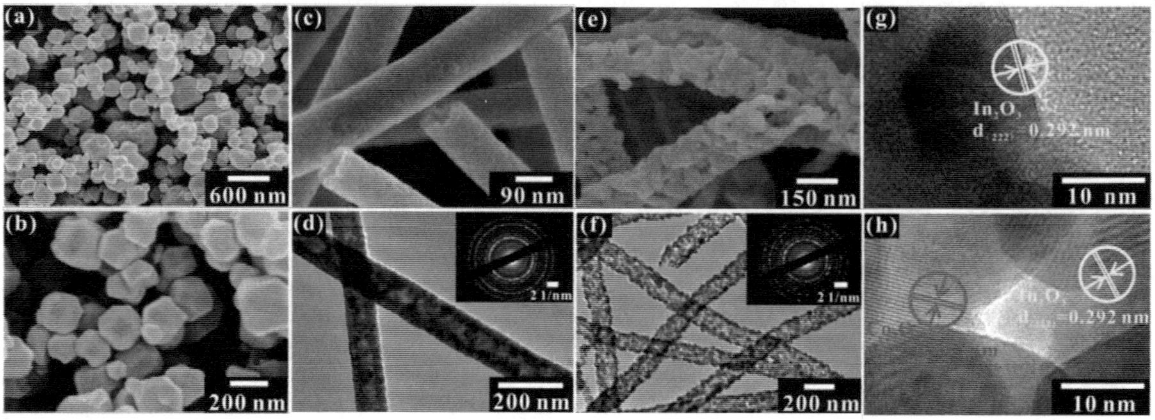


**Figure 3.** SEM images of (a, b) polyhedral ZIF-67 crystals, (c) pure $In_2O_3$ NFs and (e) 20 wt% $Co_3O_4$-$In_2O_3$ NFs; TEM images of (d) pure $In_2O_3$ NFs and (f) 20 wt% $Co_3O_4$-$In_2O_3$ NFs (The insets are the SAED patterns of pure $In_2O_3$ and 20 wt% $Co_3O_4$-$In_2O_3$ NFs, respectively); HRTEM images of (g) pure $In_2O_3$ NFs and (h) 20 wt% $Co_3O_4$-$In_2O_3$ NFs.

### 3.2 *Acetone sensing performance*

In order to investigate the gas sensing performance of $Co_3O_4$-$In_2O_3$ NFs, we monitored the samples' resistance from 200 °C to 400 °C. Subsequently, to investigate the optimum working temperature and optimum $Co_3O_4$ doping concentration, we carried out acetone sensing test of all samples to 100 ppm acetone at different working temperatures. It can be seen in Figure 4 (a) that the acetone response of pure $In_2O_3$ NFs and 10 wt% $Co_3O_4$-$In_2O_3$ NFs increases with increasing temperature. Besides, the optimal temperature of acetone response is 400 °C. In addition, the 20 wt% $Co_3O_4$-$In_2O_3$ NFs exhibited the best sensing property among them at 300 °C and the optimal temperature of 30 wt% $Co_3O_4$-$In_2O_3$ NFs is 250 °C. So the optimum working temperature and optimum $Co_3O_4$ doping concentration is 300°C and 20 wt%, respectively. The response characteristic curves of pure and 20 wt% $Co_3O_4$ loaded $In_2O_3$ NFs are shown in Figure 4 (b), it is indicated that the 20 wt% $Co_3O_4$-$In_2O_3$ NFs is much more sensitive to acetone than the unloaded one and the reproducibility of the sensing response is great. The resistance of them decreases when exposed to acetone and the acetone response for them is 38.75 and 1.67, respectively. Figure 4 (c) represents the dynamic acetone sensing characteristics of pure $In_2O_3$ NFs and 20 wt% $Co_3O_4$-$In_2O_3$ NFs at 5-100 ppm and 300 °C. The gas responses of 20 wt% $Co_3O_4$-$In_2O_3$ NFs are about 43.67, 28.99, 15.75, 6.41 and 4.17 to 100, 50, 25, 10 and 5 ppm acetone, respectively. The acetone response of 20 wt% $Co_3O_4$-$In_2O_3$ NFs (S = 43.67 @ 100 ppm) was 22 times higher compared with that of pure $In_2O_3$ NFs (S = 1.99 @ 100 ppm). Moreover, the responses of pristine $In_2O_3$ NFs toward 5-100 ppm acetone are very low (< 2) and the 20 wt% $Co_3O_4$-$In_2O_3$ NFs have a good stability. Figure 4 (d) shows that at the same working temperature, the $R_a$ of $In_2O_3$ based NFs dramatically increases with increasing $Co_3O_4$ doping content, which means that the $Co_3O_4$/$In_2O_3$ p-n heterojunctions are successfully built. $In_2O_3$ is a natural n-type semiconductor and $Co_3O_4$ is an natural p-type semiconductor with an indirect band gap of 1.6~2.2 eV. Besides, the work function of $In_2O_3$ and $Co_3O_4$ are 5.1 eV and 6.3 eV, respectively. When the $Co_3O_4$ nanoparticles interact with $In_2O_3$, electrons can transfer from $In_2O_3$ to the $Co_3O_4$ nanoparticles, and a potential barrier is formed at the $Co_3O_4$/$In_2O_3$ interface. The thickness of electron depletion layer of $In_2O_3$ NFs increases due to the forming of p-n junctions. Therefore, the resistance of $In_2O_3$ based NFs increases. In addition, the resistance of $In_2O_3$ based NFs sensors decrease with increasing temperature, which is in agreement with its semiconducting nature.

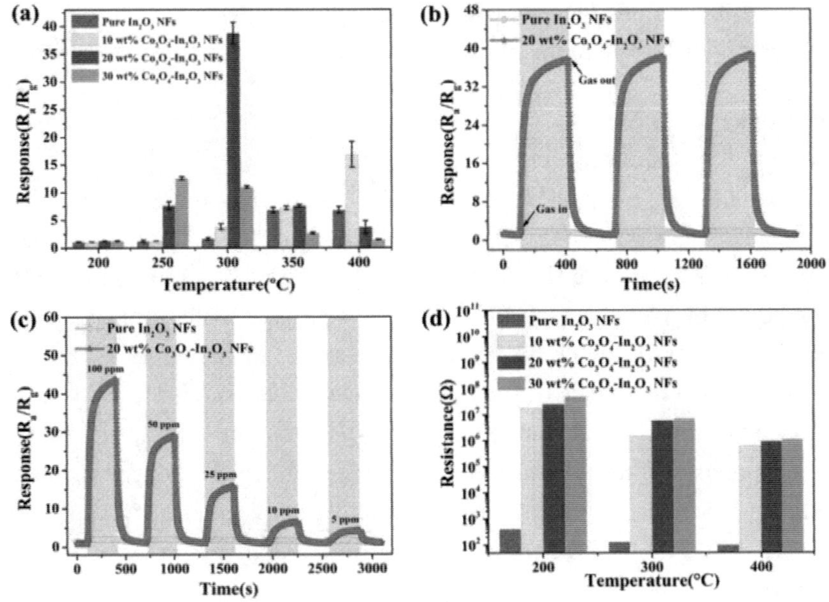


**Figure 4.** (a) Histogram plots of gas response of all samples to 100 ppm acetone at temperature from 200 °C to 400 °C, (b) the response characteristic curves of pure $In_2O_3$ and 20 wt% $Co_3O_4$-$In_2O_3$ NFs, (c) typical response-recovery characteristic of pure $In_2O_3$ and 20 wt% $Co_3O_4$-$In_2O_3$ NFs to different concentrations of acetone at 300 °C, (d) the resistance change of all samples in air at different temperatures.

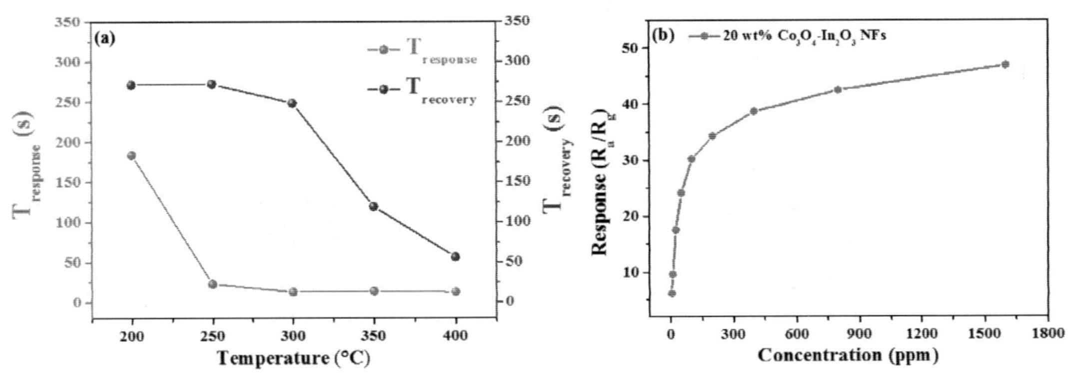


**Figure 5.** (a) Response/recovery time of 20 wt% $Co_3O_4$-$In_2O_3$ NFs to 100 ppm acetone at temperature ranging from 200 °C to 400 °C, (b) gas response of 20 wt% $Co_3O_4$-$In_2O_3$ NFs towards acetone in various concentrations at 300 °C.

Apart from gas response, the response/recovery time is also an important factor to evaluate the sensing properties of gas sensors. The response time can be defined as the time it takes from $R_g$ to $R_g$+90%*($R_a$-$R_g$) and the recovery time can be defined as the time it takes from $R_a$ to $R_a$-90%*($R_a$-$R_g$). Figure 5 (a) shows the response time of the 20 wt% $Co_3O_4$-$In_2O_3$ NFs to 100 ppm acetone decreases a lot from 183.88 s at 200 °C to 23.53 s at 250 °C and stabilizes at ~13 s over 300 °C. In addition, the recovery time keeps decreasing from 271.61 s to 54.56 s with increasing temperature from 200 °C to 400 °C.

Figure 5 (b) shows the gas response of 20 wt% $Co_3O_4$-$In_2O_3$ NFs versus acetone concentrations from 5 to 1600 ppm at 300 °C. It is found that the sensitivity increases rapidly with the acetone concentration goes up between 5 to 100 ppm, and then almost follows a linear increase with further increasing acetone concentration, indicating that the determination of acetone gas concentration is feasible.

### *3.3 Acetone sensing mechanism*

As illustrated in Figure 6, the enhanced gas sensing properties of the MOFs derived $Co_3O_4$ loaded $In_2O_3$ NFs can be explained by p-n heterojunctions theory. According to the reported literatures [5, 10], the work function and band gap of $In_2O_3$ ($Co_3O_4$) are 5.1 eV (6.3 eV) and 3.6 eV (2.07 eV), respectively. When $Co_3O_4$ is combined with $In_2O_3$, since the Fermi level of them is different, the holes transfer from p-type $Co_3O_4$ to n-type $In_2O_3$ while the electrons transfer from n-type $In_2O_3$ to p-type $Co_3O_4$. Then, an equalization of the Fermi level is formed, and the depletion layer thickness of $In_2O_3$ NFs increases and the initial resistance of $Co_3O_4$-$In_2O_3$ NFs increases significantly compared to that of pure $In_2O_3$ NFs. It is beneficial for acetone gas sensor to have a high resistance, resulting from that a small number of adsorbed acetone gas can lead to the large resistance changes during acetone sensing [11]. Furthermore, the existence of heterojunction could improve the gas sensing properties due to more active sites [12, 13]. Based on the reasons above, MOFs-derived $Co_3O_4$ in $In_2O_3$ NFs can greatly improve the acetone sensing properties.

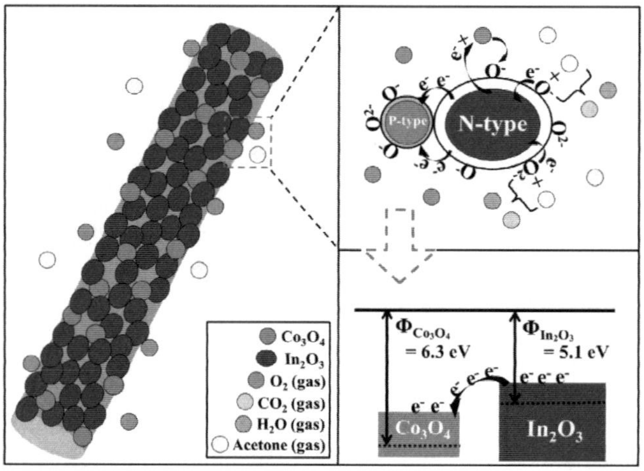


**Figure 6.** The schematic diagram of acetone sensing mechanisms of $Co_3O_4$-$In_2O_3$ NFs.

## 4. Conclusion

In summary, the MOFs-derived $Co_3O_4$ nano-grains are homogeneously dispersed on the wall of $In_2O_3$ NFs through electrospinning followed by calcination. It is found that the $In_2O_3$ based NFs consists of tiny nanoparticles strings and their average diameters are in the range from 100 nm to 200 nm. The existence of p-n ($Co_3O_4$-$In_2O_3$) heterojunctions significantly improve acetone-sensing performance. The response of 20 wt% $Co_3O_4$-$In_2O_3$ NFs (S = 43.67 @ 100 ppm) is ~22 times larger than that of pristine $In_2O_3$ NFs (S = 1.99 @ 100 ppm) at the optimal working temperature of 300 °C. These results confirm that loading MOFs-derived metal oxides in 1D metal oxide NFs through electrospinning is an efficient way to fabricate superior 1D metal oxide based gas sensors.

**Acknowledgments**

This work is supported by the Natural Science Foundation of China (11374372), the Natural Science Foundation of Shandong Province (ZR2018BEM033), Taishan Scholar Foundation (ts20130929), Fundamental Research Funds for the Central Universities (18CX02022A, 18CX02187A).

**Reference**

[1] Xiong Y, Zhu Z. Y, Ding D G, Lu W B and Xue Q Z 2018 *Appl. Surf. Sci.* **443** 114.

[2] Prim A, Pellicer E, Rossinyol E, Peiró F, Cornet A and Morante J R 2007 *Adv. Funct. Mater.* **17** 2957.

[3] Zhang C L and Yu S H 2014 *Chem. Soc. Rev.* **43** 4423.

[4] Kim H J, Jeong H M, Kim T H, Chung J H, Kang Y C and Lee J H 2014 *ACS Appl. Mater. Inter.* **6** 18197.

[5] Xing R Q, Li Q L, Xia L, Song J, Xu L, Zhang J H, Xie Y and Song H W 2015 *Nanoscale* **7** 13051.

[6] Li W H, Wu X F, Han N, Chen J Y, Qian X H, Deng Y Z, Tang W X and Chen Y F 2016 *Sens. Actuators B: Chem.* **225** 158.

[7] Xu L, Dong B, Wang Y, Bai X, Liu Q and Song H W 2010 *Sens. Actuators B: Chem.* **147** 531.

[8] Wen Z, Zhu L P, Mei W M, Hu L, Li Y G, Sun L W, Cai H and Ye Z Z 2013 *Sens. Actuators B: Chem.* **186** 172.

[9] Liang X S, Jin G X, Liu F M, Zhang X S, An S S, Ma J and Lu G Y 2015 *Ceram. Int.* **41** 13780.

[10] Jang J S, Koo W T, Choi S J and Kim I D 2017 *J. Am. Chem. Soc.* **139** 11868.

[11] Wang C, Li X, Wang B, Ma J, Cao Y, Sun Y and Lu G Y 2014 *RSC Adv.* **4** 18365.

[12] Xiong Y, Zhu Z Y, Guo T C, Li H and Xue Q Z 2018 *J. Hazard. Mater.* **353** 290.

[13] Feng C H, Li X, Ma J, Sun Y F, Wang C, Sun P, Zheng J and Lu G Y 2015 *Sens. Actuators B: Chem.* **209** 622.

IOP Conf. Series: Materials Science and Engineering **897** (2020) 012007 doi:10.1088/1757-899X/897/1/012007

# Investment Cost Analysis with Structural Design of Concrete Duct Bank Power Cables

**K Charerndee[1], R Chatthaworn[1], P Khunkitti[1], A Kruesubthaworn[1], A Siritaratiwat[1], C. Surawanitkun[2]**
[1] Department of Electrical Engineering, Faculty of Engineering, Khon Kaen University, Khon Kaen, Thailand
[2] Faculty of Applied Science and Engineering, Khon Kaen University, Nong Khai Campus, Nong Khai, Thailand
Corresponding author. E-mail: chaysu@kku.ac.th

**Abstract.** Capital investment in underground duct bank cables is very expensive. Therefore, the design and planning for underground power cables is important. It is often done with analysis of the power demands minimizing costs over the lifetime of the cables. This work presents an evaluation of underground power cable ampacity and the total costs during the anticipated operational life of the equipment. It was evaluated with variation of concrete duct bank dimensions. Analysis of ampacity depends on temperature. The thermal model of the current study uses a finite element method. The total costs evaluated by the economic model in terms of net present value analysis consist of the initial investment costs and the value of losses over the life of the cable. These were based on installation costs, load factors, annual load growth, the price of energy, the price of demand charges and the forecast discount rate. The results show that cable spacing parameters are a significant factor that influences ampacity. Ampacity can be improved while reducing the losses in power cables. This is because appropriate cable spacing can reduce the effect of Joule heating from neighboring cables. However, this leads to increased construction costs. The total costs of power cables over their lifetime can be reduced by decreasing the horizontal distance between the center of a casing pipe and the side edge of the concrete layer. This is because the contribution of reduced construction costs is more than that of the effect of increased losses. Therefore, the dimensions of the concrete duct bank significantly influence ampacity, the construction costs and the losses of power cables in an analysis of total costs over the lifetime of power cables.

## 1. Introduction

Underground power systems are widely used in urban areas because they improve the appearance of a city. Additionally, an underground power system is more secure, safe and reliable compared to overhead power systems [1-7]. However, their installation and the material costs are higher [1, 7, 8]. Maintenance and repair times are longer and it can be difficult to find the location of a cable fault [1]. Therefore, an underground cable design should consider energy demand, safety and reliability [7, 9-11].

Underground cable planning is done with the goal of installing new lines or expansions to meet the power demand over the lifetime of the cable at minimum capital costs with a sufficient level of safety and reliability. The total costs of a project consist of anticipated investment and operating costs over the system's operational life [9, 10, 12-14]. Moreover, economic evaluation is essential to project decision making.

Power cable efficiency for servicing an energy demand is specified by ampacity. This represents the maximum electrical current carrying capacity. It is limited by the critical insulation temperature [1, 8, 15, 16]. The construction conditions and material properties determine the cable ampacity. Numerical methods based on heat transfer theory are used to evaluate power cable ampacity [17-19]. The international electrotechnical commission standard, (IEC) 62095, is based on finite element analysis. It can be applied for solving more complicated problems than the conventional method [15, 16, 20-22]. Several researchers have investigated the effect of geometric configuration on power cable ampacity in a concrete duct bank and with a thermal backfill material [4, 5, 7, 15, 23-26]. In various in designs, the dimension of the duct bank is a significant factor impacting power cable ampacity.

Generally, the capital investment for cable installation increases with the structural size of a concrete duct bank. Cable ampacity is improved while their temperature is reduced under an operating current [7, 26]. Conversely, cable Joule losses decrease at lower conductor temperatures [10, 17]. The cost of such losses over the lifetime of the cable may be quite substantial. Several publications present models of economic evaluation for selection of cable sizes by minimising total costs of the project over the lifetime of the cable in terms of net present values [10-12, 20, 27]. The mathematical model proposed in the current study includes several parameters describing economic assumptions about cable operating conditions and the rate of load growth [10, 27]. Therefore, the design of the structural size of a concrete duct bank should be considered as well as the cable ampacity and the total costs over the system lifetime.

The goal of this work is to present a design with variable concrete duct bank dimension and the effect on power cable ampacity and total costs of the project over the system's lifetime. To solve this steady state heat transfer problem, a finite element method was applied. An economic model was used for evaluating the capital investment in terms of net present values. The results of this parametric analysis can be used to optimize the design of concrete duct bank power cable systems.

## 2. Models and calculations

### *2.1. Thermal model*

For analysis of underground power cable ampacity, the system voltage, $U$, was 115 kV while the frequency of a three-phase system, $f$, was 50 Hz for high voltage. The cross-sectional area of the underground power cable was 800 mm$^2$, in a pipe casing. The cable was arranged in a rectangular manner and covered by reinforced concrete in a concrete duct bank.

The power cable in the pipe casing consists of cable conductor, cross-linked polyethylene (XLPE) insulation, a metallic sheath and a non-metallic sheath or an outer jacket. There was a gap between the outer cable, inner pipe and high-density polyethylene (HDPE) pipe casing, as shown in Fig. 1. Table 1 presents the dimensions and thermal properties of a power cable. The parameters in Table 1 used for estimating the cable ampacity were adapted from [5, 28, 29]. The metallic sheath is bounded and earthed at a single point. Additionally, the gap in pipe casing is filled with a sand and bentonite mixture (SBM). The SBM can improve the cable ampacity by reducing its thermal resistivity in the vicinity of the cable [4, 5]. When the cable core temperature is higher than 90 °C, the XLPE cable insulation melts leading directly to a transmission line malfunction.

The arrangement of concrete duct bank in native soil, with the depth of the cable duct bank at $D$ = 2.5 m, is shown in Fig. 2. The parameter of $W$ represents the horizontal distance between the center of the pipe casing and the side edge of the concrete layer. The distance between the center of the pipe casing to the top and bottom of the concrete layer is defined by $H$. The parameter of $S$ is the spacing between the cable conductors.

The ambient temperature at ground level, $T_g$, was 30 °C and the native soil thermal resistivity, $\rho_s$, was 1 K·m/W [28]. The IEEE Standard 835 determined the concrete thermal resistivity, $\rho_c$, of 0.6 K·m/W [19]. The SBM thermal resistivity, $\rho_{sbm}$, was 1.0562 K·m/W [5].

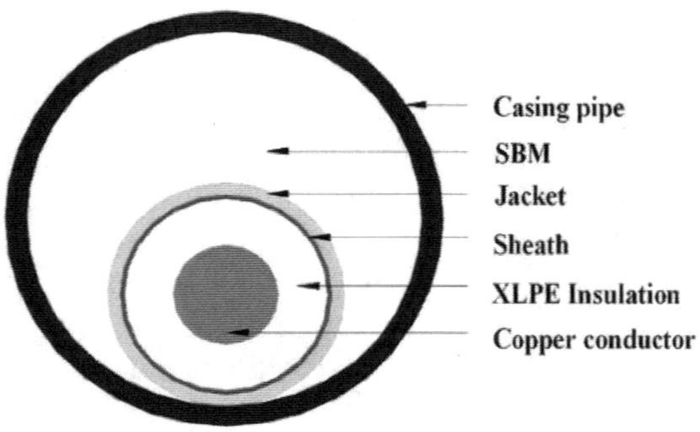


**Figure 1.** Cross-sectional layout and arrangement of a power cable in a pipe casing.

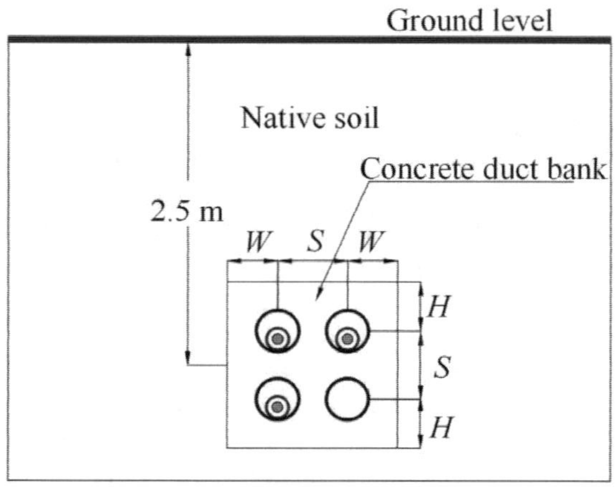


**Figure 2.** The geometry of a concrete duct bank.

*2.2. Analysis for power cable ampacity*

Power cable ampacity, $I_{90}$, calculations are required. This yields a solution of the heat transfer equations that define a functional relationship between the conductor current and the temperature of the cable and its surroundings. This was accomplished by assigning a conductor current while calculating the corresponding conductor temperature. For this analysis, the conductor current, $I$, was varied continuously with increasing temperature to the critical temperature, $T_{cc}$, of insulation, 90 °C for an XLPE insulator. The cable ampacity evaluation by FEM was slightly different from the conventional method. Moreover, the FEM can be used to solve complex problems with irregular geometries [7, 13].

**Table 1.** Dimensions and thermal properties for power distribution system at 115 kV with an 800 mm$^2$ power cable in a pipe casing adapted from [5, 28, 29].

| Description | Radius (mm) | Thermal resistivity $\rho$ (K·m/W) |
|---|---|---|
| Copper conductor | 190 | 2.5×10$^3$ |
| XLPE Insulation | 370 | 3.5 |
| Copper sheath | 385 | 2.5×10$^3$ |
| PE jacket | 430 | 3.5 |
| HDPE casing pipe | 800 | 3.5 |

The FEM was used to solve the heat conduction equation by considering steady state heat transfer in two dimensions at any point in the $xy$ plane around the power cable, calculated by solving:

$$\frac{1}{\rho}\frac{\partial^2 T}{\partial x^2}+\frac{1}{\rho}\frac{\partial^2 T}{\partial y^2}+Q=0 \tag{1}$$

where $T$ is temperature (°C) and $\rho$ is the thermal resistivity of the material (K·m/W). $Q$ is the heat generation rate in the cable dependent which depends upon the heat losses per unit of cable volume (W/(m$^3$)), defined by:

$$Q=\frac{W}{A_{hs}} \tag{2}$$

where $A_{hs}$ is the cross-sectional area of the heat source. $W$ denotes the cable heat losses per unit length (W/m). It is also called Joule losses. The heat losses of conductor $W_c$, dielectric $W_d$ and sheath $W_s$ losses (except those caused by circulating currents) can be calculated using the following equation and the material electrical properties of the IEC 60287-1-1. The cable core temperature, $T_{cc}$, was 90 °C and the cable sheath temperature, $T_{sc}$, was 70 °C [26].

The heat losses of conductor per unit length of the power cable is given by:

$$W_c = I^2 R_{c,ac}(T_{cc}) \tag{3}$$

where $I$ is the conductor current (A). It was varied to power cable ampacity, $I_{90}$. The $R_{c,ac}(T_{cc})$ is the alternating current electrical resistance of the conductor at temperature $T_{cc}$.

The dielectric heat losses per unit length of the power cable are given by:

$$W_d = \omega C \left( \frac{U}{\sqrt{3}} \right)^2 \tan \delta \tag{4}$$

where $\omega$ is the angular frequency of a three-phase system (rad/s) for an $f$ value of 50 Hz, $C$ is operating capacitance per phase and unit length, 0.227 μF/km, and $tan\ \delta$ is dielectric loss factor of 0.001.

The heat losses of the sheath per unit length of the power cable are given by:

$$W_s = W_c(\lambda_1^{'}(T_{sc}) + \lambda_1^{''}(T_{sc})) \tag{5}$$

where $\lambda'_1(T_{sc})$ is loss factor due to circulating currents at a temperature, $T_{sc}$, and $\lambda''_1(T_{sc})$ is the loss factor due to eddy currents at temperature, $T_{sc}$.

The loss due to circulating currents in the sheath is negligible [17]. In the two dimensional thermal model, it can be assumed that the right and left boundary extends to 10 m from the center of the cable duct bank. The height of the boundary from top to bottom is 10 m. The right left and bottom boundaries are assumed to be the thermal insulation. The top edge (the ground level) temperature $T_g$ was set to 30 °C [4, 5].

### *2.3. Economic analysis of the total cost of underground power cable*

Decision making in project management is done to yield the lowest total project cost. In an underground cable duct bank, the total cost for evaluation in the economic value of various dimensions of a concrete duct bank power cable is considered to include installation and operating costs during its projected operational life, expressed in terms of net present values.

The investment and operating costs have fixed and variable parts. The fixed part of the investment costs is the price of the power cable, while the variable part relates the construction work to the price of the cable. The operating costs also have fixed part and a variable part. The fixed cost is related to the costs of losses in the cable. These practically do not depend on the sizing of duct bank, while the variable part is related to the Joule losses of a power cable.

The economic value for the various dimension of concrete duct bank power cables considered the variable parts of the total project cost $CT$, calculated by:

$$CT = CI + CJ \tag{6}$$

where $CI$ is the cost of an installed length of cable duct bank, \$/m. $CJ$ is the equivalent cost at the date the installation was purchased, i.e., the net present value, of the joule losses during the anticipated operational life of $N$ years, \$/m.

#### *2.3.1. The cost for installing cable duct bank*

The installation costs of cable duct bank consist of the civil works costs and the cable cost. In this case, the cable cost is constant therefore this not comparable.

Table 2 shows typical civil engineering works for cable and their corresponding costs [26].

The civil works costs of cable duct bank can be changed by varying the design variables. Based on the data given in Table 2, a total installation cost function $CI = C\ (H,\ W,\ S,\ D)$ may be formulated as:

$$CI = 31.3\times((2\times W)+S+0.6)+47\times((2\times W)+S)\times(D+(2\times H)+S) \tag{7}$$

where *H, W, S* and *D* have the units of meters. The total costs for installation of cables was based on the research conducted by Perović in 2018 in the suburbs of the northeastern USA. However, these costs do not differ significantly from the current total installation costs for other regions of the USA [26].

**Table 2**. Costs of civil engineering that are typical for laying cables adapted from [26].

| Civil engineering work | Cost | Cost term |
|---|---|---|
| Removal of existing asphalt pavement | 13.5 $\$\cdot m^{-2}$ | $(2 \times W) + S + 0.6$ |
| Excavation and disposal of soil | 47.0 $\$\cdot m^{-3}$ | $(2 \times W + S) \times ((S / 2) + H + D)$ |
| Installation of the cable bedding (concrete cost) | 23.5 $\$\cdot m^{-3}$ | $(2 \times W + S) \times (2 \times H + S)$ |
| Repaving with asphalt | 17.8 $\$\cdot m^{-2}$ | $(2 \times W) + S + 0.6$ |

*2.3.2 The joule losses during an anticipated operational life*

To calculate the net present value of the costs and losses, it is necessary to choose appropriate values for the future development of the load, annual increases in the kW·h price and annual discount rates over the anticipated operational life of the cable. The calculation is as follows [27]:

$$CJ = \sum_{n=1}^{N} \frac{I_0^2 \times R_m \times N_p \times N_c \times ((T\times P)+D)\times\left[(1+a)^2+(1+b)\right]^{N-1}}{(1+i)^N} \tag{8}$$

where $I_0$ is maximum load in the first year, $R_m$ is the mean conductor resistance per unit length, $L$ is cable length, $N_p$ is number of phase conductors per circuit, $N_c$ is the number of circuits carrying the same type and value of load, $T$ is operating time at maximum Joule loss ($T=\mu \times 8760$), $\mu$ is load loss factor ($\mu = p\times LF+(1-p)\times LF^2$), $LF$ is the load factor and the value of the constant, $p$, can be taken as equal to 0.3 for transmission circuits. $P$ is cost of one watt-hour at a relevant voltage level \$/(W·h), $D$ is demand charge each year \$/(W·year), $a$ is the increase in the load per year, and $b$ is the increase in the cost of energy per year.

The mean conductor resistance, as an average of the values during the first and last years, is as follows [27]:

$$R_m = \frac{R_{20}}{2}\times\left(\frac{\beta+T_a}{\beta+20}\right)\times\left(\frac{1}{1-\gamma}+\frac{1}{1-g\gamma}\right)\times(1+\lambda_1) \tag{9}$$

$$\gamma = \left[\frac{I_0}{I_{90}}\right]^2 \times \left[\frac{T_{cc}-T_a}{\beta+T_{cc}}\right] \tag{10}$$

where $R_{20}$ is AC resistance of the conductor at 20 °C, $\beta$ is the reciprocal of the temperature coefficient of resistivity of the conductor material at 0 °C. For copper, $\beta$ is 234.5. $T_a$ is the ambient temperature, 30 °C, and $\lambda_1$ is sheath loss factors, $g = (1 + a)^{2\times(N-1)}$.

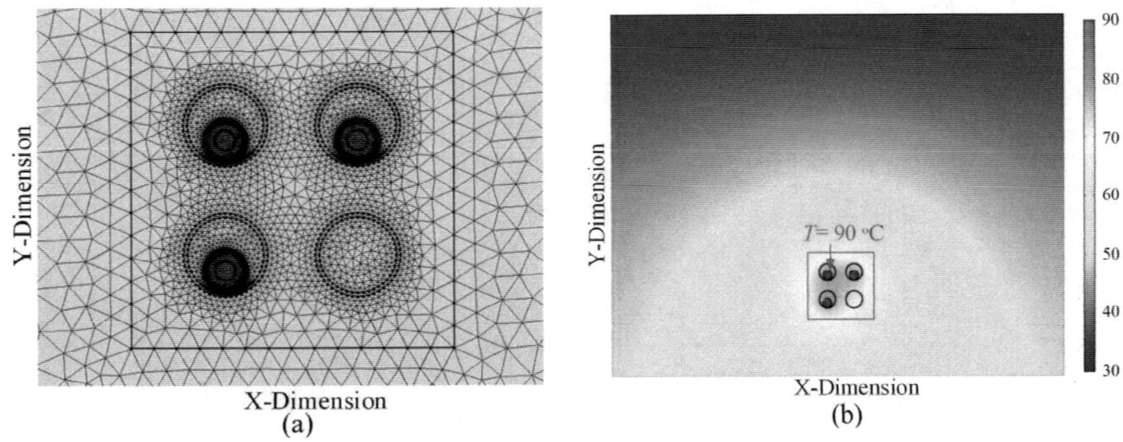


**Figure 3.** (a) The FEM mesh in duct bank power cable and (b) Thermal distribution of a traditional design using FEM.

## 3. Sample calculation

This section presents a procedure for economic analysis to arrive at the overall cost of a concrete duct bank containing underground cables over the projected operating lifetime of the line. The dimensions of the concrete duct bank system were varied to support effective decision making. The values of the design variable fell between the lower bounds, dependin on the dimensions of pipe casing pipe and concrete cover and the upper bounds, identical to those of Ocłoń et al. [4, 5]. The values of *H* and *W* were 0.09 – 0.4 m with an *S* value of 0.16 – 0.6 m.

The traditional design of a concrete duct bank has dimension of *W* equal to 0.180 m, *H* of 0.180 m and *S* of 0.250 m [29]. The results of the analysis of ampacity and the cross sectional area determined optimal values of $A_c$, of concrete corresponding to the traditional design and also to the concrete resistivity, $\rho_c$, at 0.6 K·m/W, which has an $I_{90}$ value of 902 A, *CI* of 127.04 $/m, *CJ* of 1,484.22 $/m and *CT* of 1,611.26 $/m. Fig. 3 shows the FEM mesh in a duct bank power cable and thermal distribution determined using FEM.

### *3.1. Ampacity with single design variation*

The results of determining cable ampacity at different structural sizes of concrete duct banks with *H* and *W* ranging from 0.09 – 0.4 m and *S* from 0.16 – 0.6 m with a concrete thermal resistivity of 0.6 K·m/W are shown in Fig. 4. The cable ampacity increased as the size of each concrete dimension following a polynomial function. The design of the concrete duct bank can improve the cable ampacity with varying *H* and *W* parameters from the lower to upper bound by -0.78% to 1.22% and varying the *S* parameter from its lower and upper bounds by -8.43% to 11.64% when compared to the traditional design. Additionally, the lower value of *S* parameter results in insufficient cable capacity during last year of its operational life (887.922 A). The smallest possible value of *S* is 0.23 m.

### *3.2. Costs of civil works with single design variation*

Although ampacity is important in the design of underground cable systems, construction costs of a concrete duct bank are the main point to consider in project evaluation. The construction costs of a concrete duct bank change with the dimensions of the design as is shown in Fig. 5. The construction costs of a concrete duct bank increased with the values of the *H*, *W* and *S* variables in a linear manner. The value of *H* influences the construction costs of a concrete duct bank by -4.06% to 9.93%. The rate of change was 57.34 $/m. The construction cost also increased with the value of *H* from -25.15% to 61.47%. Its rate of change was 354.94 $/m while *S* increased by -3.69% to 61.33%. Its rate of change was 218.6 \$/m compared to the traditional design. The costs are most sensitive to changes in the *W* distance than to *H* and *S*.

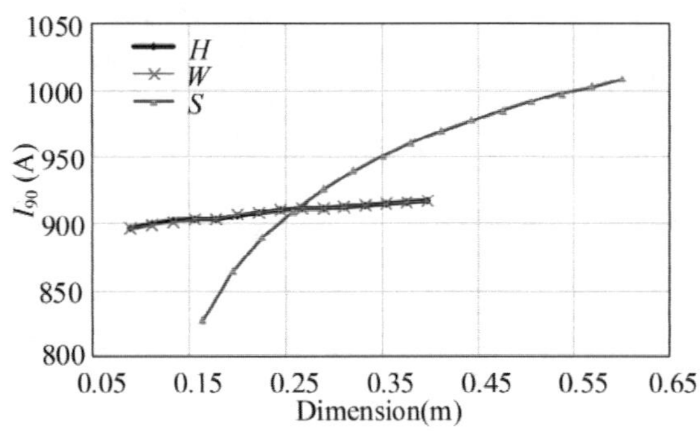


**Figure 4**. Cable ampacity, $\Delta I_{90}$ while varying the dimensions of the concrete duct bank.

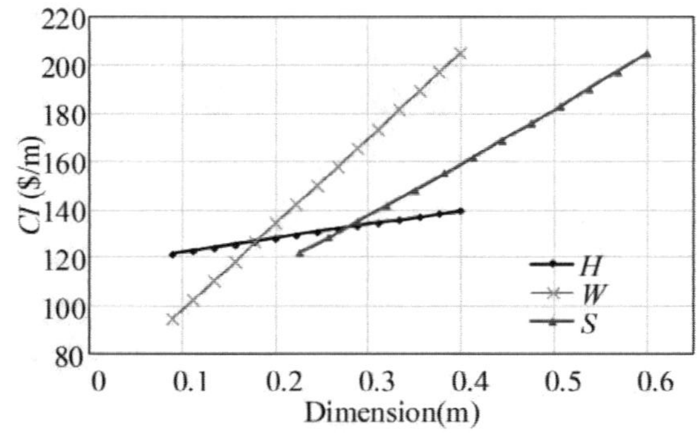


**Figure 5**. Construction costs, *CI* versus as a function of the dimension of the concrete duct bank.

*3.3 Present value of the Joule losses with single design variation*

The net present value of the Joule losses during the anticipated operational life is a significant factor in project evaluation. In this section present, the net present value of the Joule losses over 30 years with electrical and financial data are shown in Table 3. The various structural sizes of concrete duct banks are shown in Fig. 6. The Joule losses decreased with the size of each concrete dimension following a polynomial function. A concrete duct bank can reduce Joule losses by 0.23% to -0.41% by varying the *H* and *W* parameters from their lower to upper bounds. Varying *S* from its lower to upper bounds reduced losses by 0.26% to -3.27% compared to the traditional design. Moreover, varying the *S* parameter more efficiently reduced Joule losses of the cables than varying *H* and *W*.

**Table 3.** Electrical and financial data for the current case study.

| Data | Symbol | Cost term |
|---|---|---|
| Anticipated operational life | $N$ | 30 year |
| Maximum load in the first year | $I_0$ | 500 A |
| Load factor | $LF$ | 0.85 |
| Cost of energy | $P$ | $90 \times 10^{-6}$ \$·(W·h)$^{-1}$ |
| Demand charge per year | $D$ | 0.5 \$·(W·year)$^{-1}$ |
| Annual increase of load | $a$ | 2 % |
| Annual increase of cost of energy | $b$ | 7 % |
| Annual discounting rate | $i$ | 7 % |

*3.4 Total specific costs for economic evaluation with single design variation*

Transmission planning is aimed at installing new lines or expanding existing utilities to meet the power demand at minimal total costs with a sufficient level of quality and reliability. The total specific variable costs for economic evaluation in the current study involved a summation of the construction costs and the Joule losses during the anticipated operational life of the project, as is shown in Fig. 7. The value of the total specific costs when the paramter, *W*, increased followed a linear function from 1,582.70 \$/m to 1,683.32 \$/m. Changing the value of *W* produced a monotonic increase in the total specific costs due to the effects of increasing costs of the concrete duct bank. However, the results of varying the *H* and *S* values increased following a polynomial function. The lowest and highest of the total specific costs of *H* were 1,609.29 \$/m at 0.11 m and 1,617.84 \$/m at 0.40 m. Additionally, the lowest and highest of the total specific costs of *S* were 1,606.41 \$/m at 0.32 m and 1,640.71 \$/m at 0.60 m. Changing the *H* and *S* parameters had a non-monotonic response on the total specific costs.

Increasing the $H$ and $S$ parameters increased the construction costs but reduced Joule losses. Therefore, the variation function of the total costs based on changes in various parameters is a disparate function.

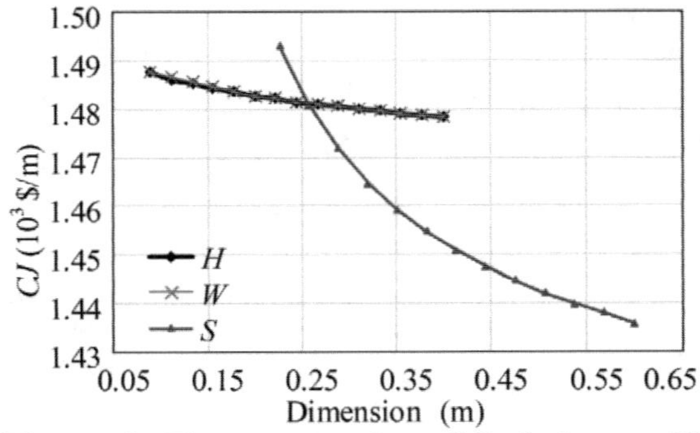


**Figure 6**. The present value of Joule losses, CJ versus each of the dimensions of the concrete duct bank.

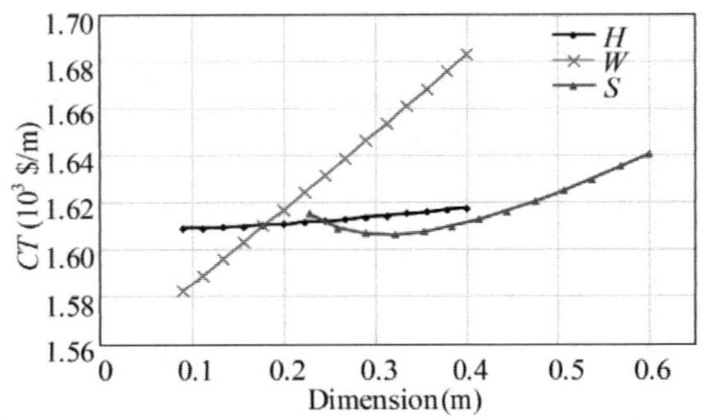


**Figure 7**. The total costs of the concrete duct bank power cable, CT versus each of the dimensions of the concrete duct bank.

For the purpose of determining the concrete duct bank dimensions for minimal total specific costs over the lifetime of a unit length of cable, the current study examined two cases with different dimensions. Case 1 defined $H = 0.18$ m, $S = 0.32$ m (the minimum total specific cost point) and varied $W$ from 0.09 – 0.4 m. In Case 2, $H = 0.18$ m, $W = 0.09$ m (the minimum total specific cost point) and $S$ was varied from 0.16 – 0.6 m. The minimum total specific costs over the lifetime of cable per unit length in both cases was the same at H = 0.18 m, W = 90 m and S = 0.32 m. This decreased the total costs by 2.09 % (1,577.57 \$/m) and improved cable ampacity by 2.99% (929 A) compared to the traditional design, as shown in Fig. 8. The concrete duct bank design for minimal total specific costs over the lifetime of cable per unit length occurred where $H = 0.11$ m, $W = 0.09$ m and $S = 0.32$ m (minimum point with single parameter variation). This decreased the total specific costs by 2.19% (1,576.00 \$/m) and improved cable ampacity by 2.55% (925 A) compared to the traditional design.

Finally, this study presents an example of a primary concrete duct bank design to minimize the total costs over its lifetime on the basis of cable length. Additional analysis is needed to determine if there will be any variation in the results due to location, environment, demand load and financial parameters.

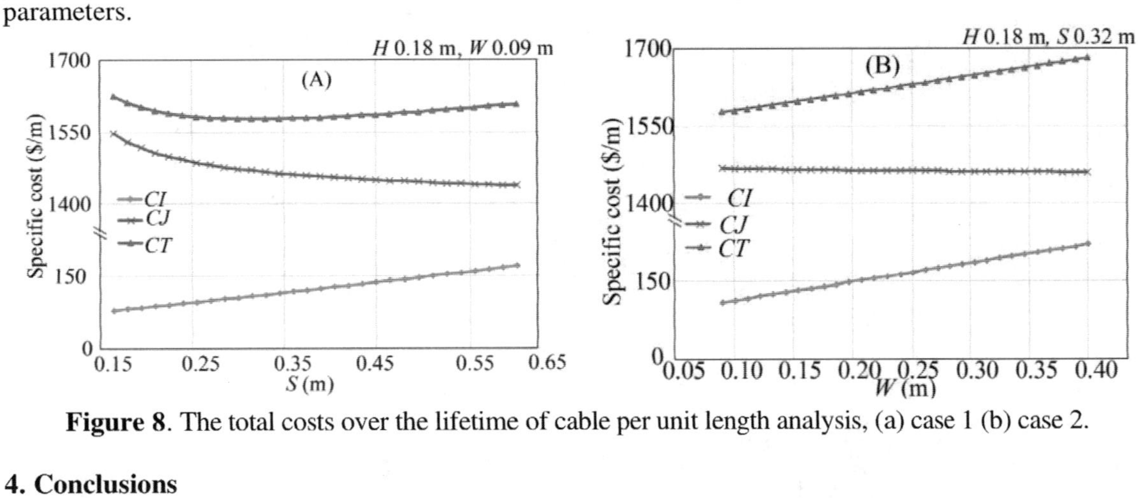


**Figure 8**. The total costs over the lifetime of cable per unit length analysis, (a) case 1 (b) case 2.

## 4. Conclusions

A parametric study of the effects of structural design of concrete duct bank power cables on cable ampacity and total specific costs over the lifetime of a power cable system is presented. The analysis includes an evaluation of cable ampacity by the FEM and measured economic values as net present values.

The sizing of a concrete duct bank substantially affects power cable ampacity and the total costs over an operational lifetime. Ampacity depends on temperature. It can be modeled using a thermal model with finite element analysis. Increased spacing between power cables, *S,* is important to improve power cable ampacity and reduce heating effects. In this case, increased spacing between power cables has a non-monotonic effect on the total costs.

Decreasing the horizontal distance between the center of the casing pipe and the side edge of the concrete layer, *W,* is important to reduce the total costs of over the operational lifetime of such systems, through reduced construction costs. However, the three dimensional parameters have different impacts on minimizing total costs. Varying three parameters during concrete duct bank design is more effective in reducing total costs and improving cable ampacity.

## Acknowledgments

This paper was based on research conducted with cooperation between the Provincial Electricity Authority (PEA) and Khon Kaen University. It was funded by PEA.

## References

[1] Bicen Y 2017 Trend adjusted lifetime monitoring of underground power cable *Electr. Pow. Syst. Res.* **143** 189-196
[2] Moutassem Wand Anders G. J 2010 Calculation of the Eddy Current and Hysteresis Losses in Sheathed Cables Inside a Steel Pipe *IEEE Trans. Power Deliv.* **25(4)** 2054-2063
[3] Orton H 2013 History of underground power cables *IEEE Electr. Insul. Mag.* **29(4)** 52-57
[4] Ocłoń P, Bittelli M, Cisek P, Kroener E, Pilarczyk M, Taler D, Rao RV and Vallati A 2016 The performance analysis of a new thermal backfill material for underground power cable system *Appl. Therm. Eng.* **108** 233-250
[5] Ocłoń P, Cisek P, Taler D, Pilarczyk M and Szwarc T 2015 Optimizing of the underground power cable bedding using momentum-type particle swarm optimization method *Energy* **92** 230-239
[6] Al-Saud MS, El-Kady MA and Findlay RD 2008 A new approach to underground cable performance assessment *Electr. Pow. Syst. Res.* **78(5)** 907-918
[7] Al-saud MS 2018 PSO of power cable performance in complex surroundings *IET. GENER. TRANSM. DIS.* **12** 2452–2461
[8] Mazzanti G 2009 The combination of electro-thermal stress, load cycling and thermal transients and its effects on the life of high voltage ac cables *IEEE Trans. Dielectr. Electr. Insul.* **16(4)** 1168-1179
[9] Kishore TS and Singal SK 2014 Optimal economic planning of power transmission lines: A review. Renew *Renewable and Sustainable Energy Reviews* **39** 949–974
[10] Anders GJ, Vainberg M, Horrocks DJ, Foty SM, Motlis J and Jarnicki J 1993 Parameters Affecting Economic Selection of Cable Sizes. *IEEE Trans. Power Deliv.* **8(4)** 1661-1667
[11] Insulated conductors committee task group 7-39, cost of losses, Loss evaluation for underground transmission and distribution cable systems 1990 *IEEE Trans. Power Deliv.* **5(4)** 1652-1659
[12] Dordevic A and Durišic Z 2018 General mathematical model for the calculation of economic cross sections of cables for wind farms collector systems *IET Renew. Power Gener.* **12(8)** 901-909
[13] Zhou H, Wu G, Pan W, Hou Y and Wang C 2013 Research on Risk Assessment of the Transformer Based on Life Cycle Cost *International Journal of Smart Grid and Clean Energy* **2(1)** 125-131
[14] Jaipradidtham C 2017 Investment cost analysis for electricity generation with renewable energy and measurement of water pressure from weir to water conservation of small hydropower project *International Journal of Smart Grid and Clean Energy* **6(1)** 67-75
[15] De León F and Anders GJ 2008 Effects of backfilling on cable ampacity analyzed with the finite element method *IEEE Trans. Power Deliv.* **23(2)** 537-543
[16] Ocłoń P, Cisek P, Taler D and Pilarczyk M 2015 Numerical simulation of heat dissipation processes in underground power cable system situated in thermal backfill and buried in a multilayered soil. *Energ. Convers. Manage.* **95** 352–370
[17] *Electric cable - Calculation of the Continuous Current Rating of Cables (100% Load Factor) and calculation of losses general.* IEC 60287 -1-1 (2014)

[18] *Electric cables – Calculations for current ratings – Finite element method*. IEC TR 62095-1 (2003)
[19] *IEEE Standard Power Cable Ampacity Tables* 2012 IEEE Standard 835
[20] Baazzim MS, Al-Saud MS and El-Kady MA 2014 Comparison of Finite-Element and IEC Methods for Cable Thermal Analysis under Various Operating Environments *Engineering and Technology International Journal of Electrical* 8(3) 470-475
[21] Dubitsky S, Greshnyakov G and Korovkin N 2016 Comparison of finite element analysis to IEC-60287 for predicting underground cable ampacity In: *IEEE Int. Energy Conf. ENERGYCON* 3-8
[22] Radosavljevi J 2014 Thermal FEM-Based Procedure for Design of Energy-Efficient Underground Cable *Humanities & Science University Journal* **10** 162-188
[23] El-Kady MA and Horrocks DJ 1985 Extended values for geometric factor of external thermal resistance of cables in duct banks *IEEE Trans. Power App. Syst.* **5(8)** 22-23
[24] El-Kady MA, Anders GA, Motlis J and HorrOcks DJ 1988 Modified values for geometric factor of external thermal resistance of cables in duct banks *IEEE Trans. Power Deliv.* **3(4)** 1303-1309
[25] Sellers SM and Black W. Z 1996 Refinements to the Neher-McGrath model for calculating the ampacity of underground cables *IEEE Trans. Power Deliv.* **11(1)** 12-20
[26] Perović BD, Tasić DS, Klimenta DO, Radosavljević JN, Jevtić MD and Milovanović MJ 2018 Optimising the thermal environment and the ampacity of underground power cables using the gravitational search algorithm *IET. Gener. Transm. Dis.* **12(2)** 423-430
[27] *Electric cable-Calculation of the current rating -Part 3-2: Sections on operating conditions – Economic optimization of power cable size*. IEC 60287-3-2 (2012).
[28] *Electric cables – calculation of the current rating – part 2-1: Thermal resistance – calculation of the thermal resistance*. IEC 60287-2-1(2015).
[29] Mekong, Subregion Harmonisation of Power Distribution Systems in the Lower. Power Cable *HANDBOOK VOL.2 Underground Cabling Design, Construction & Installation*. International Copper Association Southeast Asia Ltd; 2012.

# Third-point flexural test on concrete-filled rectangular tubular flange girders

**Jude Shahara Rosales**[1] **and Nathaniel Tarranza**[2]

[1] MS in Civil Engineering, Graduate School of Engineering, Mindanao State University – Iligan Institute of Technology, Philippines
[2] Professor, Department of Civil Engineering and Technology, College of Engineering and Technology, Mindanao State University – Iligan Institute of Technology, Philippines

Email: shahararosales@gmail.com; nathanieltarranza@yahoo.com

**Abstract.** The influence of shear stud connectors on the horizontal displacement brought about by the occurrence of shear stress at the interface between the rectangular steel tube and in-filled concrete on Concrete-filled Rectangular Tubular Flange Girders (CFRTFG) was investigated. As a composite structure, providing shear stud connectors allow stress transfer between steel and concrete which may result for the two (2) components to work as a single unit. Accordingly, a three-dimensional finite element software called SOLIDWORKS® was employed. Static linear analysis was performed using the finite element (FE) models to study the effect of shear stud connectors on the resulting stresses and slip on CFRTFGs. Results showed that the maximum slip occurred at CFRTFG without shear stud connectors while CFRTFG under full composite action had the least slip. Correspondingly, the maximum shear stress for both CFRTFG under full and partial composite action occurred at the shear stud connectors, while the maximum shear stress for CFRTFG without shear connectors occurred at the supports. Conducting finite element analysis using SOLIDWORKS® proved to be theoretically accurate as in the case of this study.

*Keywords:* tubular flange girder, composite action, horizontal slip, shear stud connectors, finite element analysis

## 1. Introduction

Concrete-filled rectangular tubular flange girder (CFRTFG) is a type of composite structure with a concrete-filled rectangular top flange and a flat bottom flange. They have the beneficial merits of steel, such as high tensile strength and ductility, and of concrete, such as high compressive strength and stiffness [1]. Its design and proposed usage is attributed to its several advantages over steel I-girders. This includes having large local buckling resistance, large torsional stiffness, and reduced web slenderness [2] [3] [4]. Being a composite structure however, uncertainties in the composite action or interaction between the steel surface and in-filled concrete serves as a limitation in its use [5].

Ensuring proper stress transfer from a component to another is needed to achieve composite action. The connection of two parts of the composite structure is of vital importance; if there is no connection, then the two parts will behave independently [6]. Lack of composite action thereof, upon load application, causes significant slip due to the occurrence of horizontal shear stress. Such slip implies that the materials are resisting the loads individually, which in turn could affect its overall structural

IOP Conf. Series: Materials Science and Engineering **897** (2020) 012008 doi:10.1088/1757-899X/897/1/012008

behaviour. In order to achieve composite action, supplemental devices such as shear stud connectors are used to allow stress transfer thus preventing slippage.

Prior to conducting actual experimental investigations, finite element analysis (FEA) have been widely practiced as it can provide approximate and valid results on understanding the behaviour of the structure (model) considered. In this study, finite element analysis using SOLIDWORKS® [7] was conducted to determine the effects of using shear stud connectors on CFRTFGs, particularly on the possible slip that may occur at the interface between the rectangular steel tube and in-filled concrete.

## 2. Methodology

### *2.1. Design Phase*

W4x13, A36 steel I-beams were considered in the design of CFRTFGs. CFRTFGs were designed in such a way that the neutral axis is located directly below where concrete is placed. This is to ensure that the entire concrete section is subject to compression under load. Figure 1 shows the cross-sectional and isometric views, respectively, while Table 1 shows the cross-sectional dimensions.

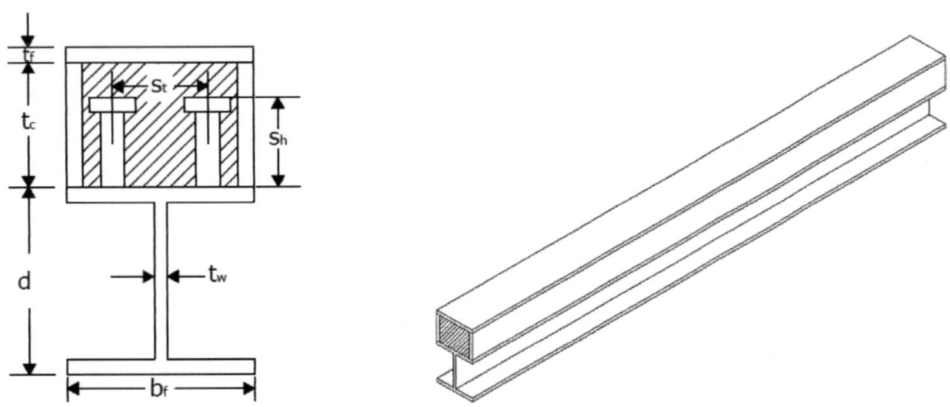


**Figure 1.** CFRTFG design: (a) cross-sectional view without concrete, and (b) isometric view

**Table 1.** CFRTFG design dimensions.

| Flange Thickness ($t_f$) | Concrete Thickness ($t_c$) | I – beam Depth (d) | Flange Width ($b_f$) | Web Thickness ($t_w$) | Shear Stud Connector Transverse Spacing ($s_t$) | Shear Stud Height ($s_h$) |
|---|---|---|---|---|---|---|
| 8.76 mm | 63.50 mm | 105.66 mm | 103.12 mm | 7.11 mm | 52 mm | 52 mm |

Three (3) models were considered: CFRTFG without shear stud connectors, CFRTFG under full composite action, and CFRTFG under partial composite action. Full composite action provides full shear connection to carry the entire section capacity, whereas partial shear connection can only carry a portion of such.

As calculated, in accordance with the specifications from the National Structural Code of the Philippines 2015, CFRTFG under full composite action have ten (10) shear connectors with a center-to-center spacing of 240 mm. The corresponding number of shear stud connectors for CFRTFG under partial composite action is fixed to be at 60% of the number provided for CFRTFG under full composite action, thereby having six (6) shear connectors with a center-to-center spacing of 400 mm as shown in Figure 2. It should be noted that in going from a full composite to a partial composite beam, the number of shear studs can be reduced significantly which also reduces the stiffness of the beam and therefore increases deflections [8].

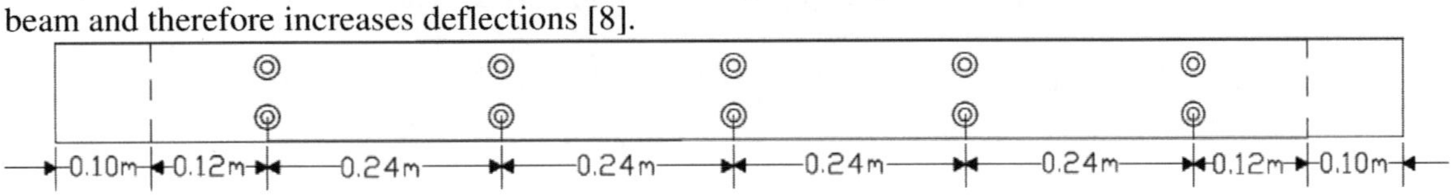


(a)

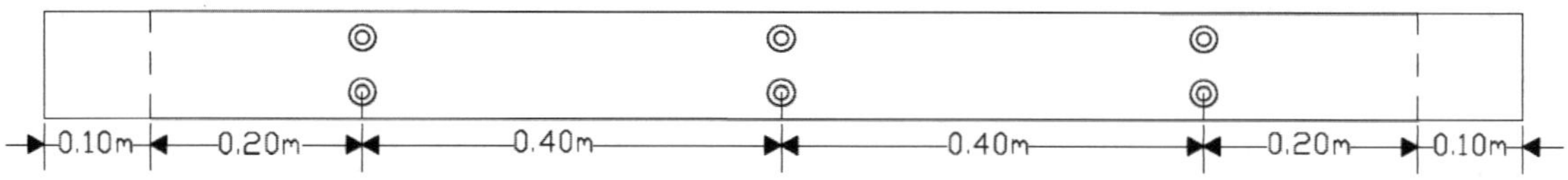


(b)

**Figure 2.** Design results for (a) CFRTFG under full composite action, and (b) CFRTFG under partial composite action

### *2.2. Finite Element Model*

Simply-supported W4x13 CFRTFGs with a length (L) of 1.40-m were developed into three-dimensional models using SOLIDWORKS®. Properties for steel, concrete, and shear stud connectors were selected using the "Materials" option in SOLIDWORKS® as shown on Table 2. Curvature-based meshing was incorporated in simulating CFRTFGs since it primarily supports both multi-core surfaces and volume meshing for complex parts or assemblies [9]. It creates more elements in higher curvatures automatically. With the development of CFRTFG in SOLIDWORKS® involving three (3) parts: steel, concrete, and shear stud connectors; applying the curvature-based meshing for simulation after assembly allows for a tighter mesh generation in complex areas (e.g. shear stud connectors) which meant for more accurate results.

**Table 2.** Material properties applied in SOLIDWORKS®.

| Material | Yield Stress (MPa) | Modulus of Elasticity (MPa) | Poisson's Ratio |
|---|---|---|---|
| Steel (A36) | 248.00 | $2.00x10^5$ | 0.26 |
| Concrete (Class AA) | 33.81 | $2.73x10^4$ | 0.20 |
| Shear Stud Connectors (AISI 1020) | $3.52x10^2$ | $2.00x10^5$ | 0.29 |

The default criterion considered was Maximum Shear Stress (Tresca) Criterion since it predicts failure to occur when the absolute maximum shear stress reaches the stress causing the material to yield in a simple tensile test [10].

### *2.3. Loading Set-up*

The simply-supported CFRTFGs are supported 100 mm from both edges. Each sample was subjected to two-point loading at the first and middle-thirds of the supported length as shown in Figure 3.

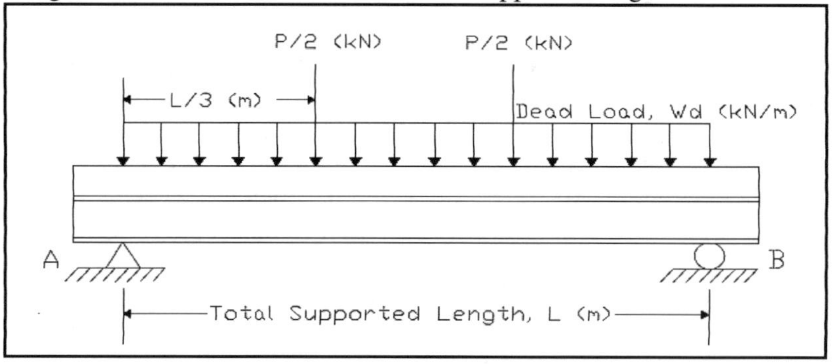


**Figure 3.** Two-point loading set-up for CFRTFGs

The load applied was the lesser of the calculated theoretical maximum capacity (P) between steel and concrete.

$$P = lesser\ between \begin{cases} F_b S_{TR(s)} = \frac{\omega_{DL} L^2}{8} + \frac{PL}{6} \\ 0.85 f'_c S_{TR(c)} n = \frac{\omega_{DL} L^2}{8} + \frac{PL}{6} \end{cases} \quad (1)$$

Where the theoretical maximum load capacity (P) is a function of the flexural stress ($F_b$), transformed section modulus ($S_{TR}$), total dead load carried by the structure ($\omega_{DL}$), and the supported length (L).

## 3. Results

### *3.1. Comparative Analysis*

Generally, the moment of inertia is a measure on the level of force that has to be applied on a certain object at a given axis of rotation; an increase on such would imply that a greater force is needed to cause rotation.

In this study, the addition of concrete-filled tubular top flange brings about an increase on the moment of inertia. This results in a reduced flexural stress on beams as well as an increase on the overall capacity, which finally minimizes deflection.

Moreover, the moment of inertia directly affects the section modulus of the structure as in Equations (2) and (3). The section modulus, in turn, is then used in calculating the capacity of the structure as shown on Equation (1).

$$I_{TR} = \left(\frac{1}{12}\right)\left(\frac{b_c}{n}\right)(t_c)^3 + (2)\left(\frac{1}{12}\right)(t_f)(t_c)^3 + \left(\frac{1}{12}\right)(b_f)(t_f)^3 + I_s \quad (2)$$

$$S_{TR} = lesser\ between \begin{cases} \frac{I_{TR}}{\bar{y}} \\ \frac{I_{TR}}{H-\bar{y}} \end{cases} \quad (3)$$

Where the transformed moment of inertia ($I_{TR}$) is a function of the dimensional properties of the structure. The transformed section modulus ($S_{TR}$) is a function of the transformed moment of inertia ($I_{TR}$), and the distance from the neutral axis to the extreme bottom fibers of the section ($\bar{y}$).

### *3.2. Effect of Shear Stud Connectors*

Theoretically, the maximum shear force and corresponding stress occurs at the supports for simply-supported structures. Thus, in obtaining values for slip between the in-filled concrete and rectangular steel tube, the maximum slip for both steel and concrete were separately obtained. Shown are the summary of results for both slip and maximum shear stress, see Table 3.

**Table 3.** Maximum relative slip and shear stress per sample.

| | CFRTFG without shear stud connectors | CFRTFG under full composite action | CFRTFG under partial composite action | Sample | Maximum Shear Stress (Pa) | Location of Maximum Shear Stress |
|---|---|---|---|---|---|---|
| Maximum Slip (mm): Concrete | $3.527 \times 10^{-4}$ | $-1.814 \times 10^{-4}$ | $1.164 \times 10^{-3}$ | CFRTFG without shear stud connectors | $7.038 \times 10^{4}$ | At 96.078 mm from the edge, near the supports |
| Maximum Slip (mm): Steel | $7.542 \times 10^{-6}$ | $7.970 \times 10^{-6}$ | $1.387 \times 10^{-3}$ | CFRTFG under full composite action | $1.398 \times 10^{5}$ | At 224.178 mm from the edge, near the end welded shear stud connectors |
| Maximum Slip (mm) between Steel and Concrete | $3.452 \times 10^{-4}$ | $-1.894 \times 10^{-4}$ | $2.230 \times 10^{-4}$ | CFRTFG under partial composite action | $1.380 \times 10^{5}$ | At 395.822 mm from the edge, near the end welded shear stud connectors |

Among the three (3) samples considered, CFRTFG under full composite action had the least slip. This is primarily due to the number of welded shear stud connectors which was designed to carry the entire section capacity. Consequently, CFRTFG without shear stud connectors had the largest slip due to the lack of means to allow shear stress transfer from one component to another.

Also, the location of maximum shear stress differs from CFRTFG with and without shear stud connectors. As shown on Figure 4, CFRTFGs under full and partial composite action experience maximum shear stress at the shear stud connectors near the supports while the maximum shear stress occurs at the supports for CFRTFG without shear stud connectors.

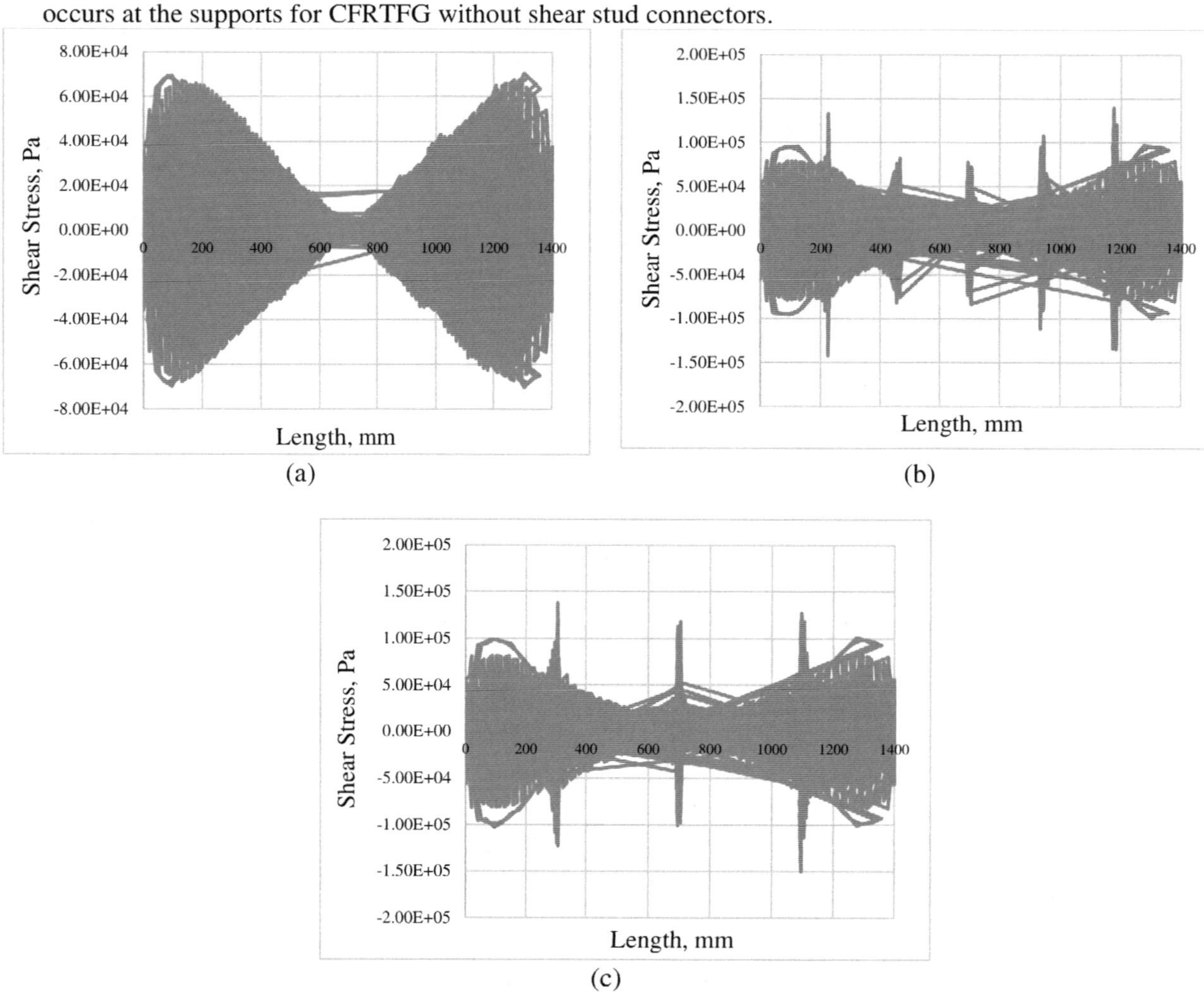


(c)

**Figure 4.** Shear stress graphs for (a) CFRTFG without shear connectors, (b) CFRTFG under full composite action, and (c) CFRTFG under partial composite action

## 4. Conclusion

In this paper, the occurrence of slippage on CFRTFGs under varying conditions was investigated. Prior to developing finite element (FE) models of CFRTFGs thru SOLIDWORKS®, the theoretical maximum capacity of 1.40-m CFRTFG, and number of shear stud connectors for CFRTFGs under full and partial composite action were calculated. CFRTFG without shear stud connectors was also considered in the study. Each model were subjected under loading located at its third-points, and were designed to fail under the Max Shear Stress (Tresca) Criterion.

Among the three (3) sample conditions considered, it was concluded that providing shear stud connectors lessens the occurrence of slip between steel and concrete. Under full composite action, where the section is designed to carry its full capacity, slip is at its lowest, while CFRTFG without shear stud connectors experienced the maximum slip. A huge decrease on the slip can be noticed upon the addition of shear stud connectors which implies that its purpose on transferring shear stress from one component (concrete) to another (steel) is effective.

In considering the occurrence of shear stress, CFRTFG under full composite action experienced the maximum shear stress at the studs near the supports, whereas CFRTFG without shear stud connectors had the least shear stress located at the supports. In this case, shear stress increases upon the addition of shear stud connectors, but it can be noticed that the maximum shear stress for the three (3) samples considered occurred near or at the supports.

It can be concluded that the finite element analysis (FEA) results was in line with the underlying principles of statics and shear stress. In the case of investigating slip, further study is needed to understand the bond stress and interaction among the three (3) materials comprising CFRTFGs. A part of proving the advantages of utilizing composite structures involve understanding its composition. Finally, in order to verify FEA results with regards to slip, experimental investigation must be conducted.

**Acknowledgment**

J S Rosales thanks Dr. Jonathan C. Maglasang, former professor at Mechanical Engineering and Engineering Sciences Department, College of Engineering and Technology, Mindanao State University – Iligan Institute of Technology, and Engr. Jeanette C. Pao for sharing their knowledge on finite element analysis using SOLIDWORKS®, and the Department of Science and Technology – Engineering Research and Development for Technology (DOST-ERDT) for making this study possible.

**References**

[1] Chen Z, Qin Y, and Wang X 2015 *Development of Connections to Concrete-Filled Rectangular Tubular Columns* Vol. 11 (Advanced Steel Construction), No.4, pp. 408-426
[2] Dong J and Sause R 2008 *Flexural Strength of Tubular Flange Girders* Vol. 65 (Elsevier) p. 622-630
[3] Sause R and Wimer M R 2004 *Concrete filled Rectangular Tubular Flange Girders with Corrugated and Flat Webs* (Lehigh University) Paper 862
[4] Wang Y and Shao Y 2018 *Stress Analysis of a New Steel-Concrete Composite I-Girder* Vol. 28 (Steel and Composite Structures) pp. 51-61
[5] Lehman, D, Roeder, C and Thody, R 2009 *Composite Action in CFT Components and Connections* Vol. 46 (Engineering Journal) p. 229-242
[6] Collings, D 2008 *Composite Construction* (ICE Manual of Bridge Engineering)
[7] SOLIDWORKS® 2018, A Product of Dassault Systèmes, Vélizy-Villacoublay, France
[8] Yura J A, Methvin E R, and Engelhardt M D 2008 *Design of Composite Steel Beams for Bridges* (Austin, Texas: Center for Transportation Research, The University of Texas at Austin)
[9] Hawk Ridge Systems 2020, Curvature Based Mesh Advantages in SOLIDWORKS® Simulation, Available at https://hawkridgesys.com/blog/curvature-based-mesh-advantages
[10] Dassault Systemes 2020 Maximum Shear Stress Criterion, Available at https://help.solidworks.com/2015/English/SolidWorks/cworks/r_Maximum_Shear_Stress_Criterion.htm

# Effect of silica fume and limestone powder on mechanical properties of ultra-high performance concrete

**P Singniao[1], M Sappakittipakorn[1,2] and P Sukontasukkul[1]**

[1] Construction and Building materials research center, Department of Civil Engineering, Faculty of Engineering, King Mongkut's University of Technology North Bangkok, Thailand
[2] Corresponding Author: manote.s@eng.kmutnb.ac.th

**Abstract.** This research aimed to study the effect of Portland cement replacement with silica fume and limestone powder in Ultra High Performance Concrete (UHPC). Nine mix proportions were designed with a constant amount of binder at 1200 kg per cubic meter of concrete. In these mixes, the cement was replaced with silica fume at 10%, 15% and 20% and limestone powder at 15% and 30% by weight of binder. For all mixes, the ratio of water to binder (W/B) and the steel fiber volume fraction were 0.2 and 2% respectively. After the preparation of test specimens and stream curing for 3 days, compressive strength, flexural strength, and length change due to shrinkage of the UHPC mixes were examined. The results showed that the silica fume increases the compressive and flexural strength (at the first peak load) but the limestone decreases. Without the limestone, the 20% silica fume mix provides the maximum average residual strength. However, when either 15% or 30% limestone is admixed, the optimum residual strength occurs at the 15% silica fume. Moreover, it is worth noting that the reducing amount of cement by replacing either with silica fume or limestone powder effectively reduces the shrinkage.

**Keywords:** Ultra High Performance Concrete, Silica Fume, Limestone Powder, Flexural Strength, Shrinkage

## 1. Introduction

Ultra-high performance concrete (UHPC) have been developed to enhance mechanical properties and to overcome the flaw of brittleness [1]. High cement content, very fine aggregate materials, chemical admixtures and fiber addition are the key factors to achieve UHPC. The excellence in its strength and toughness allow it to form thin section members without typical reinforcing steel. It is generally realized that cement manufacture generates the carbon dioxide emission. Since UHPC consumes a lot of cement, the cement replacement shall be engaged to make it more environmentally friendly [2]. Normally, silica fume is used to replace cement because it remarkably refines the UHPC microstructure [3]. Even though silica fume is a by-product collected from silicon refractory, it is not cheap. The amount of silica fume employed then shall be optimized. Limestone powder – a residue from limestone quarry – have recently been utilized as a partial replacement of cement [4-6]. The amount of limestone powder used for a cement replacement is however limited. It is interesting to increase the portion of limestone powder. Therefore, the effect of the silica fume and limestone powder combination on UHPC properties is of interest in this study.

## 2. Research Methodology

For the controlled mix, Portland cement (ASTM Type I) is used as a main cementitious binder material. The binder content is 1200 kg/m$^3$ and the water to binder ratio is kept constant at 0.20. Fine aggregates are graded #50, #100, and #200 quartz sand (i.e. passing a No.50, No.100 and No.200 sieve, respectively) [7]. The toughness of cementitious matrix is enhanced by adding hooked-end steel fiber (length of 35 mm, aspect ratio of 65, and tensile strength of 1100 MPa) at 2% by volume fraction. Polycarboxylate ether (PCE) based high range water reducing admixture is also added to achieve sufficient workability [7,8]. The proportion of controlled UHPC mix is tabulated in Table **1**.

**Table 1.** Mix proportion of controlled UHPC (kg/m$^3$)

| Mix | Binder | Quartz Sand #50 | Quartz Sand #100 | Quartz Sand #200 | Water | Fiber (2%) |
|---|---|---|---|---|---|---|
| UHPC | 1200 | 660 | 200 | 100 | 240 | 156 |

The Portland cement is partially replaced with two industrial by-products, which are limestone powder (at 0, 15, and 30% wt. of binder) and silica fume (at 10, 15, and 20% wt. of binder). Totally, there are nine mixes (UHPC1 to UHPC9 as shown in **Table 2**) investigated in this study. The limestone powder is an industrial grade refined from residues of limestone quarries. It is a very fine white powder passing a No.600 sieve. The silica fume is an ultrafine solid sphere of silicon dioxide collected as from silicon and ferrosilicon alloy furnaces. Its diameter is ranged from 0.03 to 0.3 micron. The chemical compositions of these two binders and cement are listed in **Table 3**. In addition, the physical properties and the particle size distribution of all binders and aggregates are shown in **Table 4** and **Figure 1**.

**Table 2.** Specimen codes and the variation of binder combination

| | Specimen Code* | Portland cement | Silica fume | Limestone powder |
|---|---|---|---|---|
| 1 | SF10 LP0 | 1080 | 120 | 0 |
| 2 | SF15 LP0 | 1020 | 180 | 0 |
| 3 | SF20 LP0 | 960 | 240 | 0 |
| 4 | SF10 LP15 | 900 | 120 | 180 |
| 5 | SF15 LP15 | 840 | 180 | 180 |
| 6 | SF20 LP15 | 780 | 240 | 180 |
| 7 | SF10 LP30 | 720 | 120 | 360 |
| 8 | SF15 LP30 | 660 | 180 | 360 |
| 9 | SF20 LP30 | 600 | 240 | 360 |

*SF = Silica Fume and LP = Limestone Powder

**Table 3.** Chemical compositions of binder materials (% w/w)

| Composition | Portland cement | Limestone powder | Silica fume |
|---|---|---|---|
| $CaO$ | 61.33 | - | 0.48 |
| $CaCO_3$ | - | 82.85 | - |
| $Al_2O_3$ | 6.40 | 2.39 | 1.17 |
| $SiO_2$ | 21.01 | 8.43 | 88.30 |
| $Fe_2O_3$ | 3.12 | 1.41 | 4.76 |
| $MgO$ | 3.02 | 3.28 | 0.10 |
| $SO_3$ | 2.30 | 0.10 | 1.05 |

The effect of cement replacement with silica fume and limestone powder is experimentally evaluated via compression, bending, and shrinkage tests. Following the ASTM C39, the compressive strength is measured using a 100-mm diameter and 200-mm height cylinder specimen. To examine flexural performance, the ASTM C1609 is employed on a 100x100x350 mm prism specimen. For the shrinkage, the length change of a 75x75x285 mm prism is measured as per ASTM C596.

The mixing of all mixes is done using a concrete pan mixer. They are filled and consolidated into the cylindrical and prismatic molds. Three replicates are prepared for each test method. After molding, the specimens are warped with plastic sheets to prevent water loss and are allowed to set for 24 hours. For the shrinkage test, the specimens are immediately examined. Their length is measured by using a length comparator apparatus once a day for 60 days. During the test, they are kept in air-dried condition at 55±5% RH and 23±2°C. For the other tests, the specimens are steam cured at 90°C for 72 hours and then allowed to cool down for 6 hours at room temperature prior to the testing.

**Table 4.** Physical properties of binders and aggregates used for UHPC mixes

| Property | Specific Surface (cm²/g) | Density (g/cm3) | SMD* (μm) | VMD** (μm) |
|---|---|---|---|---|
| Portland cement | 3801 | 3.15 | 5.26 | 19.75 |
| Limestone powder | 6051 | 2.73 | 3.63 | 9.75 |
| Silica fume | 11521 | 2.21 | 2.36 | 4.56 |
| Quartz sand #50 | 274 | 2.65 | 82.64 | 183.68 |
| Quartz sand #100 | 560 | 2.65 | 40.46 | 85.99 |
| Quartz sand #200 | 2176 | 2.65 | 10.4 | 34.51 |

*SMD = Sauter Mean Diameter (μm)
**VMD = Volume Mean Diameter (μm)

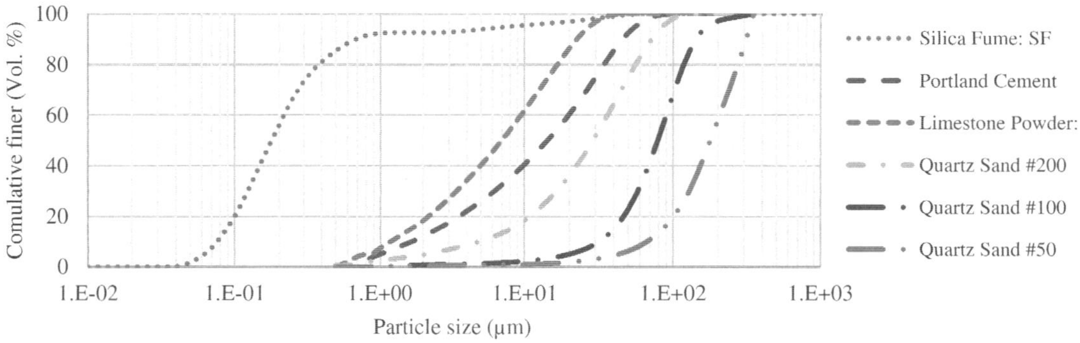


**Figure 1.** Particle size distribution of all binders and quartz sands used for UHPC mixes

## 3. Results and Discussion

### *3.1. Compressive Strength*

For compression test, the cylindrical specimens are axially loaded in a 2000-kN compression machine. The ultimate load applied is converted to compressive strength as shown in **Figure 2**. From the result of the UHPC1 to 3 (where the cement is replaced with only silica fume at 10%, 15%, and 20% w/w of binder, respectively), it is clearly apparent that the replacing cement with silica fume increases the compressive strength. That is a consequence of the filler effect from its ultrafine particles and of the additional pozzolanic reaction. Similar to the UHPC1 to 3, the group of UHPC4 to 6 and the group of UHPC7 to 9 also replace cement with silica fume at 10%, 15%, and 20%. Yet, the limestone powder is also introduced to replace cement at 15% (in UHPC4 to 6) and 30% (in UHPC7 to 9) by mass of binder. The compressive strength result of all UHPC mixes is illustrated in Figure 2. The addition of limestone powder decreases the UHPC compressive strength. The decrease is directly proportional to the added amount of the limestone powder. At this high strength level, its chemical bonding (if any) is inferior to the main matrices with silica fume. Therefore, it may be considered that the limestone powder functions like a filling material.

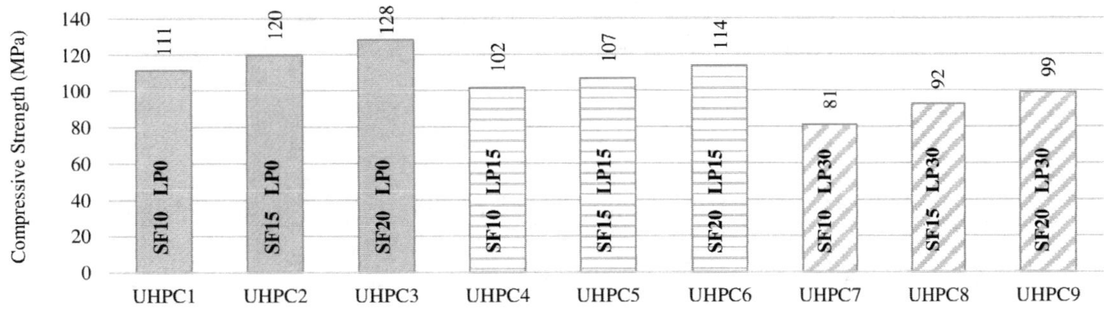


**Figure 2.** Compressive strength results of all UHPC mixes

### *3.2. Flexural Performance*

The key value of UHPC is not only the high compressive strength but the flexural performance as well. The UHPC generally provides superior flexural strength and toughness because of fiber addition and remarkable interfacial bonding between fibers and matrices. Following ASTM C1609 test method, the load vs net deflection results of all nine mixes are displayed in **Figure 3**. At the first peak load, the flexural strength of all mixes is in the range of 7.8 to 10.1 MPa as shown in **Figure 4**. It is apparently that when there is no limestone powder the 20% silica fume (the highest content) provides the maximum flexural strength. However, when the limestone powder is intermixed, the flexural strength is decreased. After the first peak load, it is assumed that the matrix phases start cracking macroscopically. The fibers then bridges the cracks and carries further load applied. The load carrying capacity after this point as shown in **Figure 3** exhibits the deflection hardening. The average residual flexural strength (ARS) is presented in **Figure 4**. When only silica fume is used (in Mix 1 to 3), the ARS is proportional to the silica fume content. It is interesting that, when the limestone powder is introduced (in Mix 4 to 9), the ARS is increased except the 20% silica fume (in Mix 6 and 9).

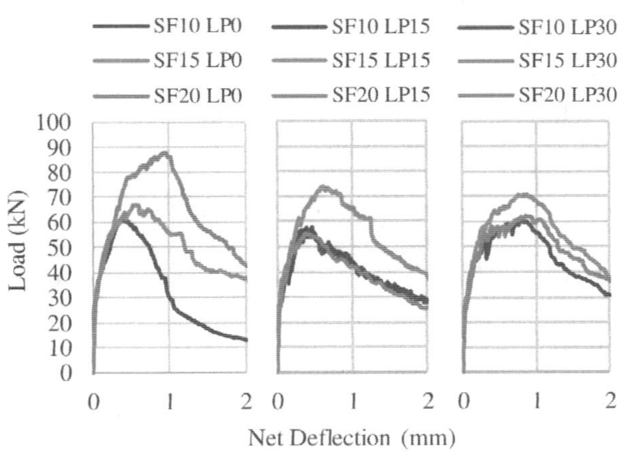


**Figure 3.** Load vs net deflection results of the UHPC mixes as per ASTM C1609 test

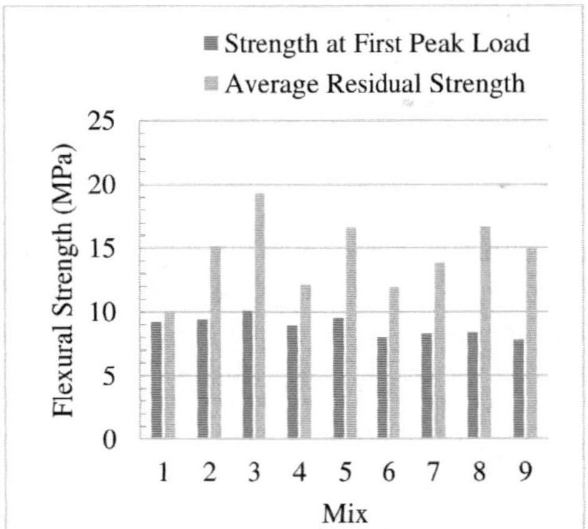


**Figure 4.** Results of flexural strength at first peak load and average residual strength

### *3.3. Shrinkage results for 60 days*

During the shrinkage test, all prism specimens are allowed to shrink freely in a temperature and humidity control room. The change of their length is measured once a day for 60 days. Due to shrinkage, the length of all specimens is shortened. **Figure 5** shows the specimen's contraction in term of shrinkage percentage. At the first couple of days, the initial shrinkage of all mixes is very high. The shrinkage continues but at a decreasing rate. During the second month of testing, the shrinkage change is very little. For a comparison among the nine mixes, their shrinkage value at the end of the test is summarized in **Figure 6**. It is noticed that while the cement is replaced with silica fume and limestone powder the shrinkage is reduced. The shrinkage is then correlated to cement content as shown in **Figure 7**.

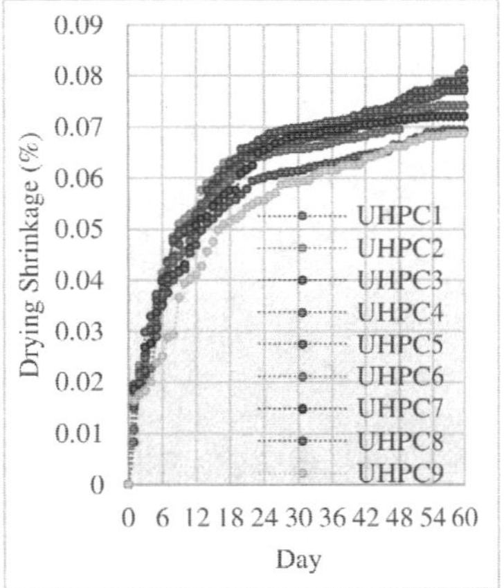


**Figure 5.** Shrinkage result of the nine UHPC mixes for 60 days

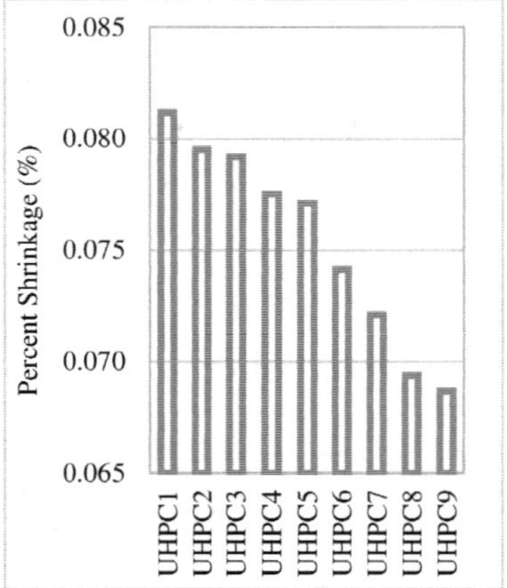


**Figure 6.** Shrinkage percentage of the UHPC mixes at the age of 60 days

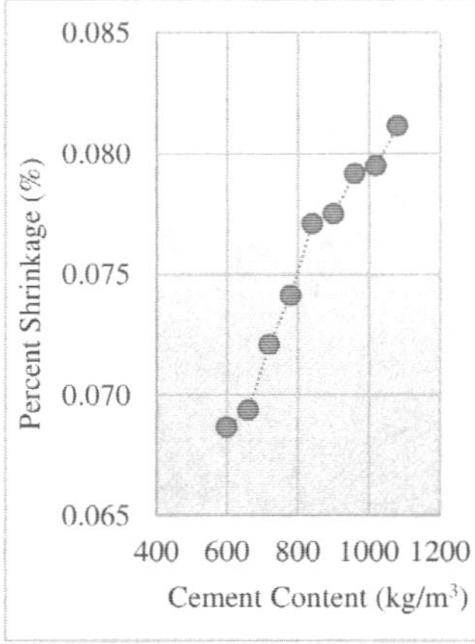


**Figure 7.** Influence of Portland cement content on the shrinkage of the UHPC mixes

## 4. Conclusions

The influence of cement replacement with silica fume and limestone powder on the compressive strength, flexural performance, and shrinkage of UHPC are assessed. According to the experimental result, conclusions can be drawn as follows. Silica fume improves the compressive strength. On the other hand, it is reduced by the addition of the limestone powder. For the flexural strength at the first peak load, the effect is similar to the compressive strength. Without the limestone powder, the 20% silica fume yields the maximum average residual flexural strength. When either 15% or 30% limestone powder is admixed, the average residual strength is optimum with the 15% silica fume (in mix 5 and 8). The cement content significantly affect the shrinkage. As the cement is reduced by the replacement with silica fume and limestone powder, the shrinkage is clearly decreased.

## Acknowledgements

The authors would like to thank the Siam Research and Innovation Company Limited for the particle size evaluation of the materials used in this study. We also thank the SR Fiber Company Limited for providing the steel fiber and the Siam City Concrete Company Limited for providing the Portland cement and chemical admixtures.

## References

[1] Yoo DY and Banthia N 2016 *Cem Concr Compos* **73** pp 267-280
[2] Soliman NA and Tagnit-Hamou A 2017 *Constr Build Mater* **139** pp 374–83
[3] Lee NK, Koh KT, Kim MO and Ryu GS 2018 *Cem Concr Res* **104** pp 68–79
[4] Li W, Huang Z, Cao F, Sun Z and Shah SP 2015 *Constr Build Mater* **95** pp 366–74
[5] Camiletti J, Soliman AM and Nehdi ML 2013 *Mater Struct Constr* **46(6)** pp 881–98
[6] Huang W, Kazemi-Kamyab H, Sun W and Scrivener K 2017 *Cem Concr Compos* **77** pp 86–101
[7] Courtial M, De Noirfontaine MN, Dunstetter F, Signes-Frehel M, Mounanga P, Cherkaoui K, et al. *Constr Build Mater* **44** pp 699–705
[8] Plank J, Schroefl C, Gruber M, Lesti M and Sieber R 2009 *J Adv Concr Technol [Internet]* **7(1)** pp 5–12

IOP Conf. Series: Materials Science and Engineering **897** (2020) 012010 doi:10.1088/1757-899X/897/1/012010

# Tension analysis of bolted sleeve connection between pultruded GFRP tube and steel member considering adhesion

**Zhihua Xiong[1], Yuqing Liu[2], Zhengzhong Wang[1*], Yun Sun[2]**

[1]College of Water Resources and Architectural Engineering, Northwest A&F University, Yangling 712100, China
[2]Department of Bridge Engineering, Tongji University, Shanghai 200092, China
[*]Email: wangzz0910@163.com

**Abstract.** Pultruded Glass fibre reinforced polymer (GFRP)-steel hybrid structures have become increasingly popular in civil infrastructures. However, the connection between GFRP and steel members is one of the key issues limiting the development of the innovative structure. This paper presents tensile capacity of bolted sleeve connection between GFRP and steel members. With consideration of the pre-loaded bolt effect, the appropriate fastening torque was proposed by finite element investigation. The adhesive effect was also considered and investigated by cohesive zone method. The bolted sleeve connection provides a good solution for the tubular and large-span structures.

## 1. Introduction

Pultruded Glass fibre reinforced polymer profile, which has the advantage of high strength to weight ratio, good durability performance, ease of installation and lower life-cycle cost, has been attracting an increasing interest in the field of civil engineering. Hybrid steel and GFRP structure is a common practice in bridges [1,2], space frame [3] and other infrastructure [4]. For each application, connections are inevitable to make the two kinds of materials action together.

The connection between steel and GFRP could be classified as adhesive and mechanical joints. Keller conducted a series of static and fatigue experiments on adhesive connections finding that composite action between FRP bridge decks and steel girders can be expected and no damage was observed in the bond during fatigue tests [5-6]. The mechanical joints can be various, including clamps, bolts and shear studs, among which bolts are most common in FRP and steel connection. Du Peloux [3] made ultimate tests on GFRP tubes connected by pinned steel sleeve when design the gridshell of Ephemeral Cathedral in Paris as shown in Figure 1. Chakherlou found that with the clamping force increasing, the fatigue strength of FRP plates increase, but the fatigue life only improved at a low stress level [7]. Davalos investigated a non-grouted sleeve-type connection for stiffness, strength, fatigue resistance, and degree of composite action [8]. Satasivam used a novel blind bolt on steel-FRP composite beam systems and studied the effect of bolt rows and pultrusion orientation on the joints' stiffness and capacity [9]. Dakhel studied energy absorption and load carrying capacity of steel-FRP connections in shear wall systems with various factors, for example bolt numbers [10]. Shahryarifard proposed a term named lamination cohesion rate [11], which was used to determine the failure mode of steel tubes connection by GFRP sleeve as shown in Figure 2.

In column-beam connection and space frame's connection, hybrid steel and GFRP structures also have wide applications [12]. Ascione tested GFRP beam-column connection with adhesive [13], which was found that the adhesive beam-column connection presented reliable performance as well as bolted connection. Shahkhosravi [14] carried out fatigue experiment of GFRP laminates with drilled hole, which involved the delamination effect of drilling and presented an evident bending fatigue improvement of drilling hole in lamina. Experimental analysis on octatube steel joint of GFRP footbridge was made and it's said that the joint's capacity showed no decrease under 2.1 million fatigue load cycling [15]. Van Wingerde made a comparison between common bolt and bolt with epoxy under different fatigue ratio [16].

To the author's knowledge, the understanding about tensile behaviour of the pultruded GFRP tubular member and its connection are limited. Especially, GFRP tubular member with hole easily took place stress concentration on the edge of the pre-loaded bolt's hole. Encompassing the adhesive layer in the bolted sleeve and the pre-loaded bolt, finite element (FE) models were built and analysed to investigate the tensile performance of the connection.

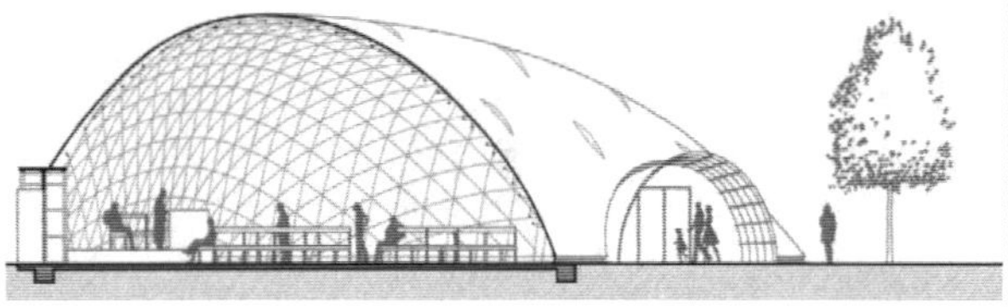

**Figure 1.** Gridshell of Ephemeral Cathedral in Paris [3]

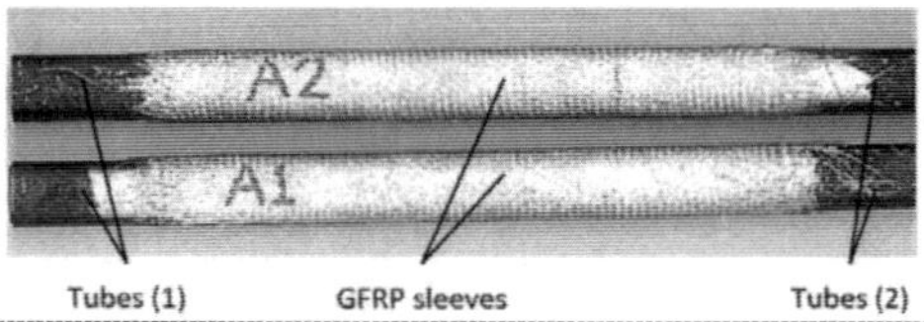


**Figure 2.** Steel tubes with GFRP sleeve [11]

## 2. Bolted sleeve joining GFRP tube and steel member

In tubular structures, pultruded GFRP tube could make full use of its light-weight material potential for a tendon as shown in Figure 1. In Figure 3, the GFRP tube's end was fastened by bolt and adhesion, and the steel sleeve was welded to the gusset plate connecting to the chord. In space frames and bridges, the main skeleton made up of pultruded GFRP could be extended by the configuration as shown in Figure 4 [17]. After the tube was inserted in the sleeve, bolt was fastened to the prescribed torque with the existing epoxy layer on the tube. The GFRP tube has a relative low compressive strength than steel. The pre-loaded bolt yields stress concentration on the edge of the GFRP tube's hole. Next, a connection was built by FE. The radius of the sleeve and GFRP tube was 36mm and 30mm respectively, the thickness of GFRP tube was 6mm the same as that of steel sleeve, the bolt was M12 in accordance with Chinese code.

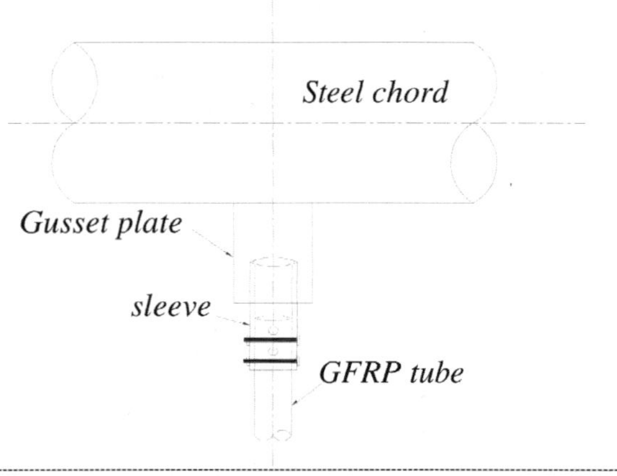


**Figure 3.** Bolted sleeve joining steel member and GFRP tube

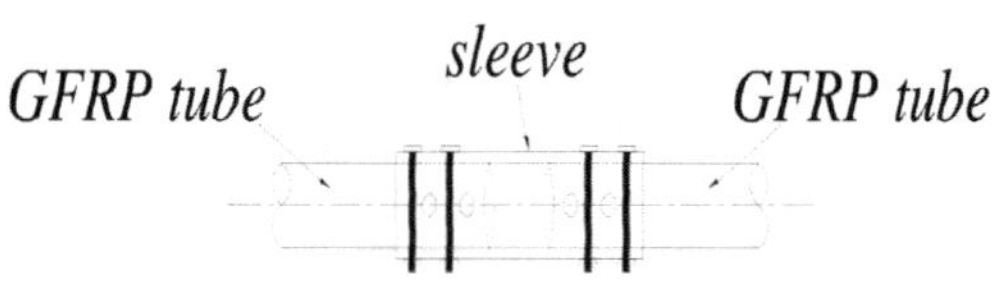


**Figure 4.** Bolted sleeve joining GFRP tube

## 3. FE model of bolted sleeve

The FE software Abaqus was employed to simulate the tensile behavior. GFRP material property was assumed to be anisotropic which required seven input parameters-Young's modulus of roving direction ($E_1$=26GPa) and transverse direction ($E_2$=16GPa), shear modulus ($G_{12}$=1.25GPa, $G_{13}$=1.25GPa, $G_{23}$=15.8GPa), Poisson's ratio ($v_{12}$=0.3, $v_{23}$=0.3). A bilinear material constitution was assumed for steel in which Young modulus and yield stress were defined as $2.1\times10^5$MPa and 345MPa respectively. Bolt's material property adopted the same model as steel while the yield strength was 1043MPa [12]. All the models were meshed by reduced integration solid element C3D8r. The contact pairs were divided into two types. One is the standard penalty contact for GFRP tube, sleeve and bolt. The other is adhesive layer. Cohesive zone method (CZM) has been proved to be an efficient way to simulate the adhesive behavior [18]. When implementing CZM, maximum interfacial bond stress or separation needs to be determined. In this paper maximum interfacial bond stress was assumed to be 0.05MPa [19]. The cohesive law was illustrated in Figure 5, in which the adhesive traction degraded as the separation developed. It evolved under the assumption of the fracture energy $G_c$ of cohesive behavior.

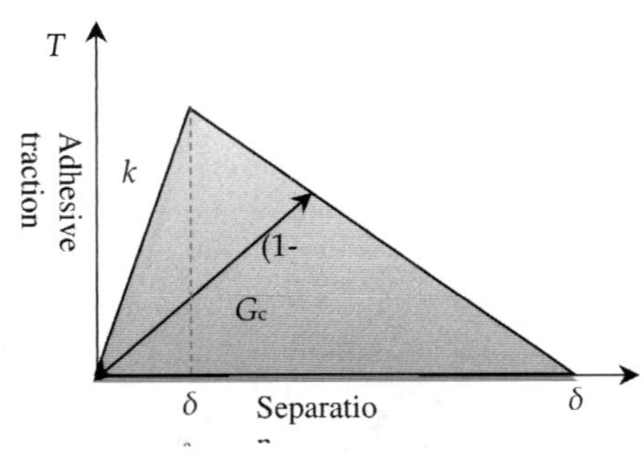


**Figure 5.** Adhesive constitution

The bottom surface of the sleeve was fixed as shown in Figure 6. There were two load steps in FE, the first step is bolt fastening and the second is tension.

Bolts were designed to be fastened to three kinds of torques in FE, which were 2.4Nm, 4.8Nm and 10 Nm. Pre-tension load was applied by bolt load in the software. FE models are listed in Table 1.

**Table 1.** Specimens' detail

| Specimens | Bolt Number | Bolt's torque(Nm) |
|---|---|---|
| BSC1 | 2 | 2.4 |
| BSC2 | 2 | 4.8 |
| BSC3 | 2 | 10.0 |
| BSC4 | 3 | 4.8 |

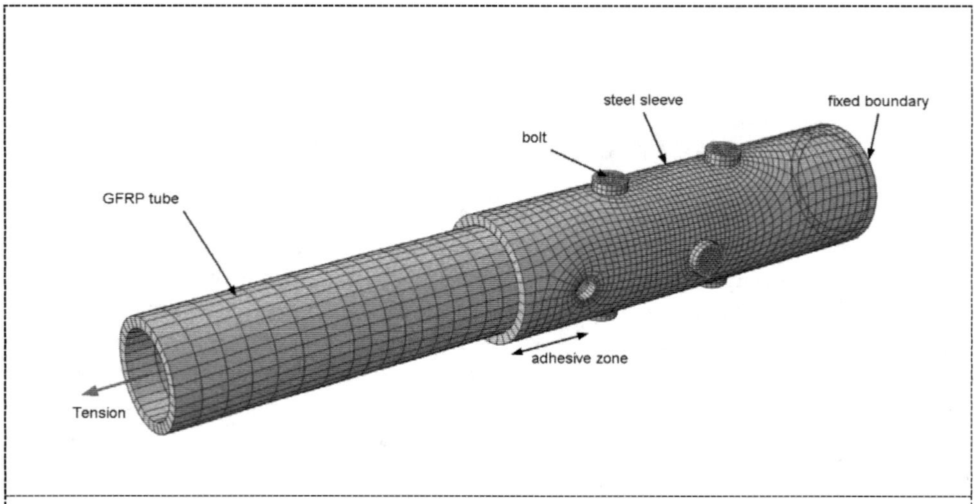


**Figure 6.** FE model of bolted sleeve connection

## 4. Analysis results

All the specimens reached its peak load when the GFRP tube's bolt hole failed due to compression, and the failure mode of specimens was GFRP tube's hole bearing failure as shown in Figure 7. Typical load-displacement curves are plotted in Figure 8. The ultimate strength of the sleeve was determined by the number of the bolt as observed from Figure 8, in which the ultimate load of BSC4 with three bolts was much larger than that of specimens BSC2 with two bolts. Adhesion layer didn't make significant effect since the smooth surface and the limited interfacial stress. With respect to the bolt's torque, specimens BSC2 hole's edge compressive stress due to fastening torque was 5.2 MPa, while BSC3's stress was nearly 11MPa.

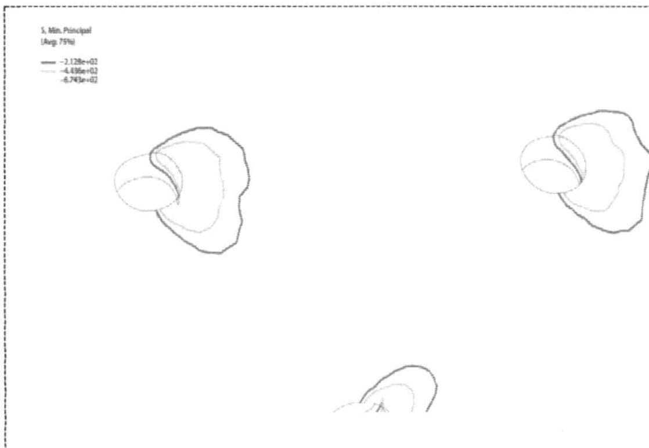

**Figure 7.** Failure mode of bolted sleeve

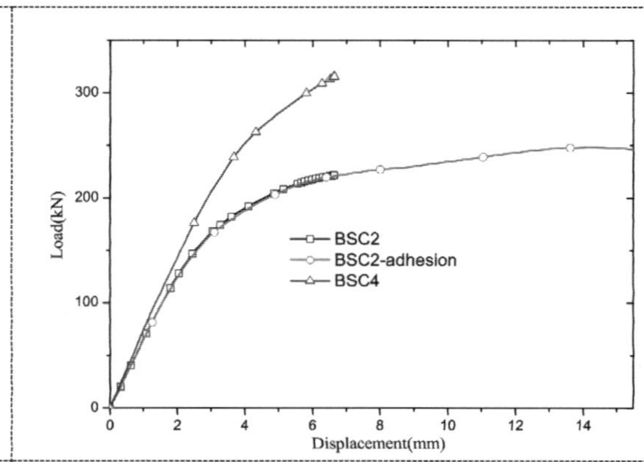


**Figure 8.** Load-displacement curves

## 5. Conclusions

This paper presented a tension analysis of bolted sleeve joining GFRP tube and steel members. Incorporating the bolt's pre-tension and adhesion in the numerical simulations, the following conclusions were drawn. The tensile strength of bolted sleeve is governed by the compressive strength of GFRP tube and the number of bolt. Maximum bolt fastening torque of M12 is suggested to set to 4.8N.m. The adhesive layer of the bolted sleeve doesn't show a substantial effect on the ultimate tensile strength of the connection. In the future work, the fatigue performance of the bolted sleeve connection between GFRP tube and steel member needs to be further evaluated.

**Acknowledgments**
The authors acknowledge for the financial support from National Natural Science Foundation of China, grant number 51908422.

**References**
[1] Nelson M, and Fam A 2014 *Constr. Build. Mater.* **50**, 368-376.
[2] Xiong Z, Liu Y, Zuo Y and Xin H 2019 *Compos. Struct.* **222**, 110938.
[3] Du Peloux L, Tayeb F, Baverel O and Caron J 2016 *Struct. Eng. Int.* **26**(2), 160-167.
[4] Yu T, Hu Y and Teng J 2016 *J. Constr. Steel Res.* **124,** 12-22.
[5] Zhou A, Keller T 2005 *Compos. Struct.* **69**(3), 336-345.
[6] Keller T, Guertler H 2005 *Compos. Struct.* **70**(4), 484-496.
[7] Chakherlou T, Oskouei R, Vogwell J 2008 *Eng. Fail. Anal.* **15**(5), 563-574.
[8] Davalos J, Chen A, Zou B 2011 *J. Compos. Constr.* **15**(3), 441-450.
[9] Satasivam S, Feng P, Bai Y, Caprani C 2017 *Eng. Struct.* **138**(MAY1), 63-73.
[10] Dakhel M, Donchev T, Hadavinia H 2019 *Thin-Walled Struct.* **134**, 52-60.
[11] Shahryarifard M, Golzar M, Safarabadi M 2018 *Int. J. Adhes. Adhes.***82**, 79-89
[12] Zhang Z, Wu C, Nie X, Bai Y and Zhu L 2016 *Compos. Struct.* **157**, 51-61.
[13] Ascione F, Lamberti M, Razaqpur A and Spadea S 2017 *Compos. Struct.* **160**: 1248-1257.
[14] Shahkhosravi N, Yousefi J, Najafabadi M, Burvill C and Minak G 2019 *Compos. B: Eng.* **163**: 536-547.
[15] Yang X, Bai Y, Luo F, Zhao, X and Ding F 2016 *Compos. Struct.* **138**: 227-236.
[16] Van Wingerde A, van Delft D and Knudsen E 2003 *Plastics, rubber and compos.* **32**(2): 71-76.
[17] Luo F, Bai Y, Yang X and Lu Y 2015 *J. Compos. Constr.* **20**(1), 04015024.
[18] Xiong Z, Liu Y, Zuo Y and Xin H 2019 *Mater.* **12**(12), 1906.
[19] Yang J, Smith S and Feng P 2013 *Thin-Walled Struct.* **62**, 215-228.

# Mechanical Properties and Sustainability of Palm Kernel Shell Powder (PKSP) as a Partial Replacement of Cement

**Gunalaan Vasudevan* and Maneswaran Subramaniam**

Department of Construction Management, Faculty of Built Environment, Tunku Abdul Rahman University College Malaysia

*Email: gunalaanv@tarc.edu.my

**Abstract.** This research showed the results of experiments evaluating the use of palm kernel shell powder from palm oil refinery waste production as partial replacement of Ordinary Portland Cement (OPC). Many researchers have studied the use of agro-waste as constituents in concrete but not as a partial replacement of cement specifically. Therefore, the objective of this research is to identify the performance of palm kernel shell powder as partial replacement of cement in the production of concrete. Palm kernel shell powder in various amounts, namely 5%,10%,15% and 20 % by volume was added as a replacement for Ordinary Portland Cement (OPC). The results showed that palm kernel shell powder concrete greatly improved the workability, compressive and flexural strength of concrete. All of the testing were followed the American Society for Testing and Materials (ASTM). The workability of concrete was tested by using slump test to check the consistency of freshly made concrete. For compressive strength, a total of 27 cubes with size 150mm x 150mm x 150mm were used to determine the compressive strength of concrete when replace with 5%, 10%,15% and 20% of palm kernel shell powder as a replacement of cement in concrete. The results showed that palm kernel shell powder concrete greatly improved the compressive and flexural strength of concrete. The rate of water absorption of palm kernel shell powder reduced as palm kernel shell powder filled up the existing voids while making it impermeable. However, the compressive strength of the palm kernel shell powder decreases gradually when the amount increased. It can be concluded that the optimum amount of palm kernel shell powder as partial cement replacement is 15%. In the correction, an experimental investigation of ultrasonic pulse velocity(UPV), carbonation test and rebound hammer test was undertaken to palm kernel shell powder and admixtures as partial replacement cement in concrete.

**Keyword:** Palm kernel shell powder, admixture, and Mechanical Properties

## 1. Introduction

Sustainable development of construction industry should be carried out in order to protect the future generation. [1] By-products and agro-waste materials such as bottom ash, crushed concrete fines, filter residue, paper ash, palm kernel shell and lignite fly ash predicted to have potential for use in building materials, for instance as raw materials for clinker production, as supplementary cementitious materials (SCMs) or mineral additions in concrete. The properties of the products are dependent on the reactivity of the waste materials used, which can be classified as inert, latent hydraulic or pozzolanic. [2] In this study, waste materials were first characterised. This was followed by theoretical considerations of the mentioned application options.

IOP Conf. Series: Materials Science and Engineering **897** (2020) 012011 doi:10.1088/1757-899X/897/1/012011

Oil palm is genuinely a golden crop of Malaysia. Oil palm is developed for its oils. As vegetable oil seed crop, the oil palm is a proficient converter of sunlight based vitality into biomass. [3] Other than being a productive maker palm and kernel oil, it additionally creates various residues and by products. The by-products of oil palm industry are from the field and factory. Palm kernel shells are one of the wastes from palm oil industry, which have for quite some time been utilized as fuel in boiler to create steam and power for plant forms [4]. Palm part shell is the hard shell of the oil palm organic product seed that is broken to take out the piece utilized for extricating palm oil. Thus, it is the by-products of palm oil processing during which the palm oil is extracted.

## 2. Problem Statement

First and foremost, it is expected that the usage of Ordinary Portland Cement (OPC) will be increased by 100 trillion tons by 2050 due to rising of infrastructural development all over the world. [5] Around 900 kg of $CO_2$ is released per ton of concrete conveyed by current practices. In like manner, the OPC is assessed to have contributed 5–7% of worldwide anthropogenic $CO_2$ emanations in 2009. [6] This carbon dioxide emission leads to one of the main cause of climate changes all over the world. For an example, acid rains will occur and it may make the infrastructure like buildings to deteriorate easily due to the chemical substances that form from carbon emission. [7] Therefore it is essential to find a solution to find a suitable material that can be replaced in the cement, so effect of carbon emissions can be reduced. Meanwhile, oil palm factories in Malaysia have continued extending as a result of the degrees of fame of human needs, for instance, vegetable oil. Palm kernel shell is one of wastes produce at oil palm industrial facilities. In Malaysia, in excess of 4 million tons of bit palm shell has been made yearly [8].

## 3.Methodology

In this research is to determine the performance of the concrete that contain the palm kernel shell powder by preparing concrete cube sample and tested to obtain some of the basic engineering properties. The concrete mix design is done by systematic analysis and chooses the proportion of the ingredient to use the concrete mix to produce an economical concrete and also with strength that desired when the cube is hardened. The variables which can be controlled are water cement ratio, maximum aggregate size, aggregate grading and use of admixture.

| **Phase 1** | To identify palm kernel shell powder as addition into concrete in percentage | Remarks |
|---|---|---|
| **Material Selection** | | **If no, back to material** |
| | Identify Tests/Experiment to be carry out | selection |
| | | |
| | Analysis of Particle size distribution and Properties of palm kernel shell powder | |
| | | |
| **Phase 2** | Mixture of materials and percentage of palm kernel shell powder as addition | **If no, back to analysis** |
| **Design and conduct** | | |
| | Mixing, casting, and compaction | |
| | Dry Shrinkage Concrete | |
| | Testing of Hardened Samples | |
| **Phase 3** | | |

| | | |
|---|---|---|
| **Long Term Behaviors** | Durability Test | |
| **Phase 4** | Application (Brick interlocking) | |
| **Performance** | | |
| | End | |

**Figure 1.** Conceptual framework

## 4.Result and Discussion

*4.1. Workability*

Figure 2 shows that the 20% complete replacement result is harsh which gave a shear slump. It is recorded very low slump (0–4 mm) meaning very low workability. Lightweight concrete of 15 N/mm2 gave slump in the range of 0–260 mm. It can be said that concrete with low slump doesn't necessarily denote high compressive strength; It was observed that slump value of 105 mm (high workability) was achieved with incorporation of small percentage of super plasticizer. Replacement of PKSP increases the slump remains a true slump which means the workability rate is high but when it increases to 20% the slump drops these is merely because the chemical properties of PKSP has overtaken the water/cement ratio rate when it loses the workability.

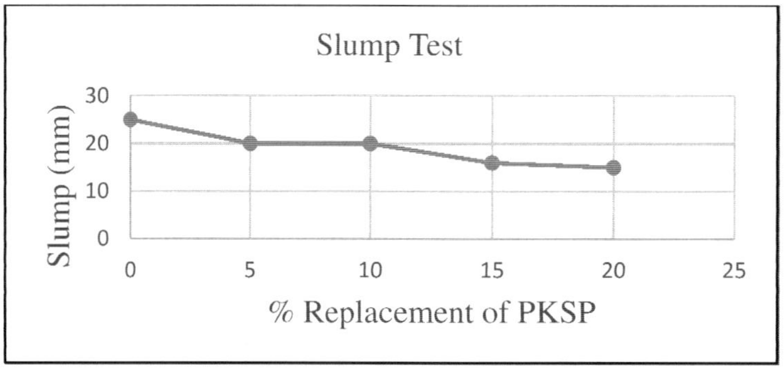


**Figure 2.** Slump test Different Percentages of palm kernel shell

*4.2. Compressive Strength Test*

Based on the Figure 3 compressive strength of all mixes increased steadily with respect to curing age. At 5% partial replacement of cement, compressive strength of concrete specimen was 26.41 MPa, 29.33 MPa and 30.21 MPa according to their age of curing days which is 7 days, 14 days and 28 days. While that of 20% were very low compared to the former results.

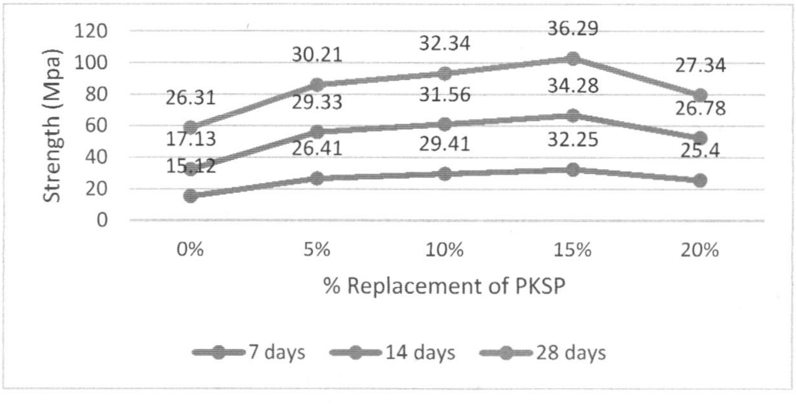


**Figure 3.** Compressive strength test

It undergoes deterioration at 20% replacement of PKSP most probably because of bond between particles is to a large extent dependent on surface texture, cement paste bonds more adequately with rough surface. Surface area of contact between the palm kernel shells is increased as the percentage of palm kernel replacement increases Since the content of cement in the concrete remained constant, the needed extra bonding was lacking leading to reduction in compressive strength. So at 15% of replacement, its shows a great compressive strength.

*4.3. Ultrasonic pulse velocity (UPV)*

**Table 1.** Results of UPV with varying % replacement of PKSP

| UPV, Direct Method | | | | | |
|---|---|---|---|---|---|
| | **0%** | **5%** | **10%** | **15%** | **20%** |
| **7 Days** | 3.412 | 4.171 | 4.678 | 4.717 | 4.212 |
| **14 Days** | 3.563 | 4.493 | 4.671 | 4.807 | 4.314 |
| **28 Days** | 3.745 | 4.610 | 4.967 | 4.912 | 4.414 |

Based on the Table 1, UPV test shows very excellent trend lines when percentage replacement of PKSP increase from 5% to 15% where the concrete quality shows good strength compared to percentage replacement of 20%. This is because at 20% of PKSP replacement, the concrete may have loss its strength maybe because there is more moisture content of concrete which leads to more porosity in the concrete which finally makes the concrete weaken with high percentage of PKSP replacement. In conclusion, higher velocities indicate good quality and continuity of the material which also means the PKSP replacement showing great improvement within the casted concrete in between 5% to 15%.

*4.4. Flexural strength test*

The test results are presented in Table 2 the flexural strength values of concrete with palm kernel shell ash (PKSP) when compare with the control concrete were increased by 5% to 20%. The flexural strength at 5% of partial replacement of PKSP was 2.7 MPa,2.9 MPa and 3.0 MPa for 7,14 and 28 days. Increases of 15% in flexural strength value of concrete contain PKSP was observed when compare with the control mix. Flexural tensile strength was increased with age which means PKSP having a good impact on the strength of concrete. Flexural tensile strength was reduced with increase in 20 % of PKSP replacement according to the data and this is possibly because of moisture condition of the concrete specimen.

**Table 2.** Results of compressive strength test with varying % replacement of PKSP

**Flexural Strength, MPa**

| | 0% | 5% | 10% | 15% | 20% |
|---|---|---|---|---|---|
| **7 days** | 2.1 | 2.7 | 2.9 | 3 | 2.6 |
| **14 days** | 2.4 | 2.9 | 3 | 3.1 | 2.7 |
| **28 days** | 2.6 | 3 | 3.10 | 3.2 | 2.8 |

*4.5. Rebound Hammer Concrete Test*

According to Figure 4 above rebound hammer test shows strength increasing trend lines where percentage replacement of PKSP increase from 5% to 15% where the concrete quality shows good strength compared to percentage replacement of 20%. This is because at 20% of PKSP replacement, the concrete may have started to loss its strength maybe because the surface and moisture condition of the concrete which finally makes the concrete weaken with high percentage of PKSP replacement. These moisture content could be adapted from the PKSP and surrounding due to age of concrete exposed to outside condition. In conclusion, uniformity of the concrete of concrete is greatly increase when the the optimum percentage reached which is 15%.

IOP Conf. Series: Materials Science and Engineering **897** (2020) 012011 doi:10.1088/1757-899X/897/1/012011

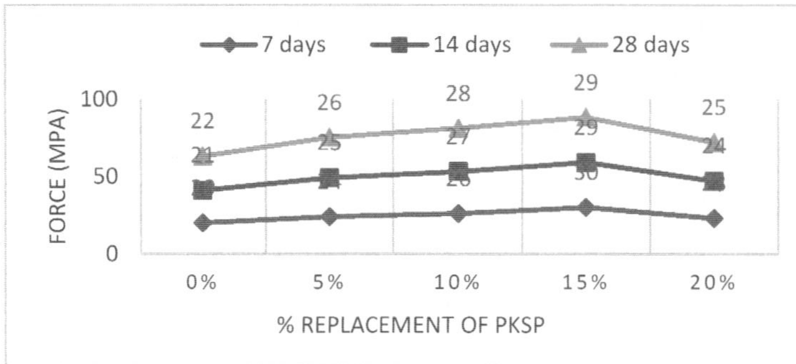


**Figure 4.** Rebound hammer test

*4.6. Microstructures of Palm Kernel Shell Powder*

The apparent morphologies of the Palm Kernel Shell Powder examined by SEM are shown in Figure 5(a) it was observed that C-S-H gel was vastly spread on the mixture of hydrated cement paste which was the main cause for the effective strength. Formation of Portlandite $Ca(OH)_2$ and Calcite ($CaCO_3$) was visualized all over the exterior surface of hydrated cement paste. These expansion and distribution of mineral elements was one of the reasons for the effective strength of the mix. Replacement of fine aggregate with manufactured sand did not exposed any flaw in strength but, the range of distribution of minerals was changed due to replacement of fine aggregate with manufactured sand. The SEM image in Figure 5(b) SEM micrograph of the harden cement paste, the distribution of C-S-H was nearly decrease at stage of 28 days due to replacement of cement with PKSP. In this mix, the range of development of C-S-H was less which may be due to unreacted particles present in the mix.[9]

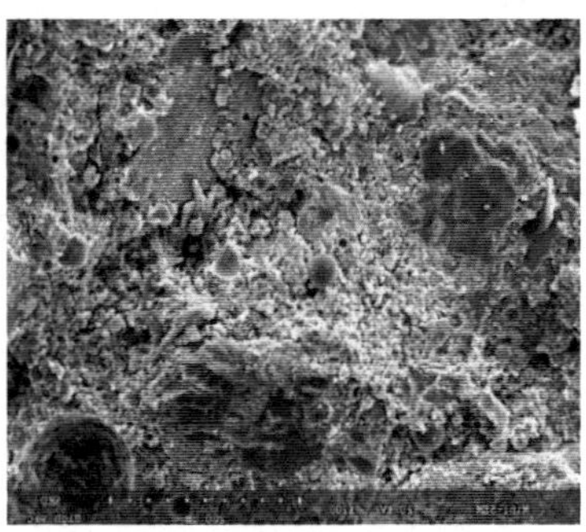

**Figure 5(a).** SEM images of control mix

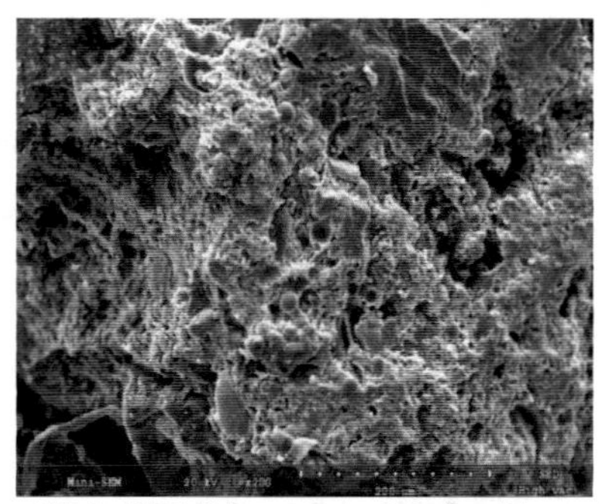

**Figure 5(b).** SEM images of Palm Kernel Shell Powder

*4.7. FTIR spectra and TGA analysis*

Based on the results from Figure 5(a) IR spectra palm kernel shell powder was determined by transform infrared (FTIR) spectrometry of treated palm kernel shell powder respectively. The presence of characteristic band of -O-N = N- group at 1,031.8 cm–1 and C-O stretching at the region of 1,300 to 1,000 $cm^{-1}$. Figure 5(b) shows result from TGA showed that there were two stages of weight losses. For the first stage, minor weight loss occurred at temperature below 200 °C - 500°C attributed to an adsorption of water molecules to the surface of the material and loss of organic compound.

IOP Conf. Series: Materials Science and Engineering **897** (2020) 012011 doi:10.1088/1757-899X/897/1/012011

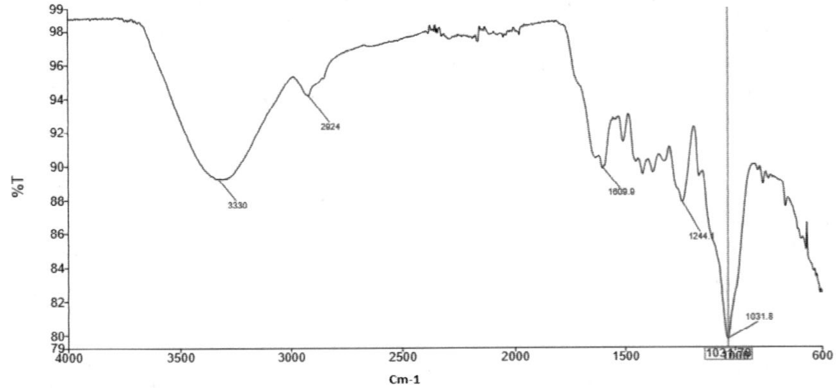


**Figure 5(a).** FTIR spectra for raw Palm Kernel Shell Powder

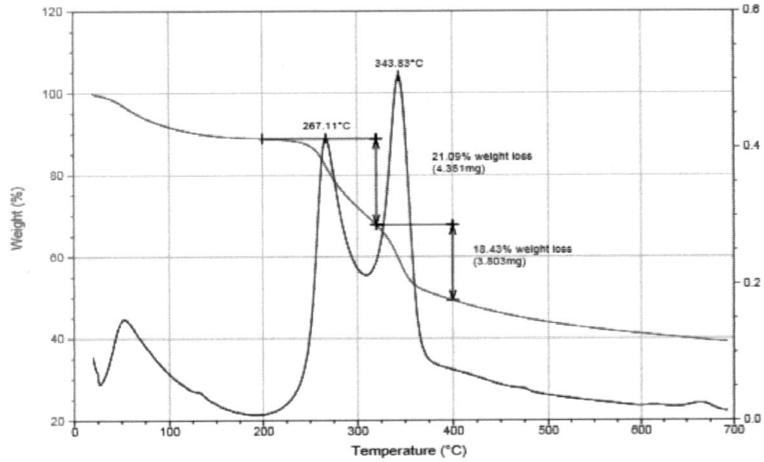


**Figure 5(b).** TGA analysis for Palm Kernel Shell Powder

## 5. Conclusion and Recommendation

The overall objective of the work was to investigate the feasibility of palm kernel shell powder as a partial replacement pf cement to concrete mix. This study included the preparation of concrete mixes containing and the evaluation of palm kernel shell powder concrete properties in fresh and hardened states. The studying properties involved mix workability, compressive strength and flexural strength. According to experimental results, the usage of in palm kernel shell powder concrete mixes as an alternative of disposal for palm kernel shell powder is possible. The current type of palm kernel shell powder accumulated in dumping sites and the expected future type were used in making concrete mixes. The influence of both types on concrete properties was studied. In all cases there was an optimum quantity of palm kernel shell powder which can be used without introducing any change in mix preparation and acceptable properties were still be produced [10]. The study showed that the palm kernel shell powder can be used in production of cement without changing the normal industrial process. Recommendation for further research The following recommendations are proposed for further research and study in order to from a complete picture of using palm kernel shell powder in concrete mixes:

1) Investigation is needed on the different superplasticizer used together with palm kernel shell powder in the concrete mix design.
2) Study is required for palm kernel shell powder mix with other by-products in the concrete mix design.
3) Further investigation of the higher percentage of replacement of the cement by the palm kernel shell powder in concrete mix design.

4) Longer curing time is required in order to understand the behaviour of the concrete, and it might be giving the different result for oil palm fibre concrete.
5) Durability test such as sulphate attack test must be further study in order to understand percentage loss in weight of the palm kernel shell powder concrete by sulphate attack

## Reference

[1] A.P.C. Chan, W. Y. D. W. M. Y. D. C., 2012. Determining an optimal recovery time for sustainable environment. Building and Environment, Volume 58, pp. 163-171.
[2] A., U. I., 2012. Characterization of pozzolanic. INTERNATIONAL JOURNAL OF ENVIRONMENTAL SCIENCES, 3(1), pp. 518-524.
[3] Anon., 2019.Oil palm cultivation in malaysia. [Online] Available at: https://www.mpams.co.uk/wiki/construction [Accessed 28 March 2019].
[4] Ayemba, D., 2018. Construction Review Online. [Online] Available at: https://constructionreviewonline.com/2017/08/land-surveying-importance/[Accessed 26 December 2018].
[5] Benz, M., 2019. Ordinary Portland Cement. [Online] Available at: https://www.HOPC.com/blog/10-project-opc-that-endanger-your-projects-success [Accessed 1 July 2019].
[6] Blessen Skariah Thomas, S. K. H. S. A., 2017. Sustainable environment– A review. Renewable and Sustainable Energy Reviews, Volume 80, pp. 550-561.
[7] Chia-Wen Liao, T.-L. C., 2015. carbon emissions. Safety Science, Volume 72, pp. 363-370.
[8] D.K. Chua, Y. G., 2004. Incident causation model for oil palm knowledge. Journal of Oil Palm Industry, 130(4), pp. 542-551.
[9] Olafusi, Oladipupo & Olutoge, Festus. (2012). Strength Properties of Corn Cob Ash Concrete. Journal of Emerging Trends in Engineering and Applied Sciences. 3. 297-301.
[10] Vasudevan, G(2020) Performance of Steel Slag as a Partial Replacement Fine Aggregate Incorporating Superplasticizer. Lecture Notes in Civil Engineering Volume 59, 2020, Pages 151-160.

IOP Conf. Series: Materials Science and Engineering **897** (2020) 012012 doi:10.1088/1757-899X/897/1/012012

# Determination of Hydraulic Conductivity Value by using Sunjoto Shape Factor

**Thinzar Aye, S.Sunjoto, Djoko Luknanto**

Department of Civil and Environmental Engineering, Gadjah Mada University, Yogyakarta, Indonesia

Email: thinzaraye.geol@googlemail.com, sunysunyoto@ugm.ac.id, luknanto@365.ugm.ac.id

**Abstract.** Hydraulic conductivity, *K* value, is important for soil properties determination and there are many equations and method that can be used in this region. In 1951, Hvorslev introduced the shape factor method to determine *K* value that only depends on geometry of well. The geometry of well or shape factor is mainly focused on the length, screen and base of the groundwater well. Later, many researchers used this method and developed different kind of equations for different geometry of wells. Sunjoto also developed equations and shape factor to determine the *K* value. This method can be used easily by using iteration method. In this research, the comparison has been performed using constant laboratory test method and Sunjoto method. The experiment test was conducted in the laboratory and it took measurements for three tests. The Sunjoto equation was used for analysis to the experiment test measurements. The hydraulic conductivity value with constant head test is 0.00038 m/sec and with Sunjoto shape factor method is 0.000184 m/sec for test (1), 0.000180 m/sec for test (2) and 0.000119 for test (3). According to this comparison value, the hydraulic conductivity value is more acceptable when it uses shape factor method since the constant laboratory test is using disturb sample. Moreover, according to the result, Sunjoto method can be used as a new method to determine hydraulic conductivity value.

## 1. Introduction

Hvorslev (1951) first introduced shape factor equations to solve the error of hydrostatic time lag. These equations were developed to determine hydraulic conductivity value by focusing only on the dimension of well area [1]. Following Hvorslev, many scientists have been developed shape factor equations. Many arguments arise about several shape factor equations assumptions [2][3]. Sunjoto developed the equations in unsteady state conditions based on Hvorslev (1951). He modified this equation to compute the drawdown and coefficient of permeability[4][5][6]. In 2002, Sunjoto developed several shape factor formulas as the equation parameters. His method is well known in Indonesia and many researchers have conducted research about this equation and method [7][8]. For example, Runtu Kexia G.A., Sunjoto S., and Hendrayana H. had conducted the research using the Sunjoto equations for comparison with Cooper-Jacob method under pump test [9]. These researches and papers mentioned that Sunjoto equation can be used easily under the pump test. Most of the previous equations, which were based on well formula geometry, were developed for slug test, such as the research of Hvorslev (1951), Cooper et al (1967), and Bouwer and Rice (1971) despite many researches have been used Sunjoto shape factor equation in pump test. Therefore, the researcher tries to prove clearly about kinds of more appropriate well test for Sunjoto equation. The pump test or slug test, which was performed with different kinds of wells, is not easily taken at the field area. According to this reason the researcher has decided to use experiment tank and well in the Laboratory to do the pump test and slug test. The main part of research usually has been divided into two parts; data analysis under pump test and data analysis under slug test. In this current research, the researcher has chosen to use new shape factor that had been developed by Sunjoto with appropriate well conditions under pump test.

## 2. Objectives and Methods

The main objective of this research is to prove and decide what kind of aquifer test is more suitable with Sunjoto method to determine the hydraulic conductivity value. The research method includes three parts. In the first part, the researcher had to collect data and researches that were related to the Sunjoto equation and various kinds of well condition geometry. The experiment tank had to be prepared for aquifer demonstration and for the pump test implementation. The grain size analysis test had been done in Laboratory to know the soil materials characteristics. Then, in the second part of research, the constant head test was conducted at the Laboratory and the hydraulic conductivity value from this test had been taken as reference value. In the third part, the researcher tried to perform data analysis using pump test equation. The groundwater drawdown data measurements were obtained by the experiment tank. The experiment test had been conducted for 8 hours for the first test, 12 hours for the second test and more than 14 hours for third test. The analysis was performed later, and it will be done using the well geometry to determine hydraulic conductivity value. The comparison had been done using these three results.

### *2.1. Characteristics of Soils*

The grain size analysis and constant head test were conducted at Laboratory to determine the soils characteristics (Figure 1). According to grain size distribution curve, the Hazen's effective size of D10 gravel pack is 0.123 mm and the slotted size must be equal to the D10 gravel pack. $D_{10}$ is the diameter of 10% material grain size [10]. The soil materials that have been used in this test can be determined as fine sand according to the grain size analysis test. The measurement result has been described in Figure 1.

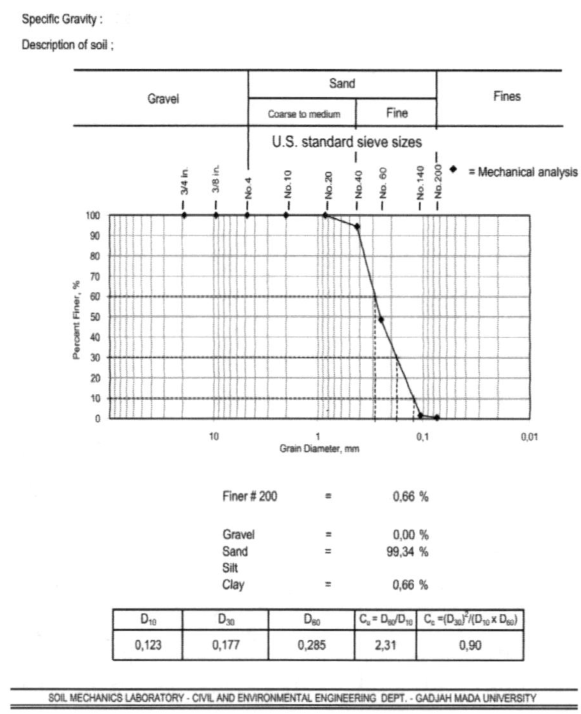


**Figure 1.** Grain Size Analysis Curve

*2.2. Determining hydraulic conductivity value by constant head test*

The constant head permeability test is a laboratory experiment to determine the soil permeability. The measurement data can use the following equation in order to get the permeability coefficient, *k* [11]. In this research, the permeability was corrected into 20°C (68°F) by multiplying k using water viscosity ratio at temperature test to the water viscosity at 20°C (68°F). The diameter of specimen is 7.5 cm, area is 44.18 $cm^2$, and volume is 220.9 $cm^3$. The average discharge $q$ is 4.495$cm^3$/sec, Hydraulic gradient $I$ value is 2.2 and the Correction factor $R_t$ is 0.829. The Hydraulic Conductivity $K$ value is 3.8E-0.4 m/sec based on the constant head test at the Laboratory.

**Table 1.** Measurement Data of Constant Head Test

| t (sec) | Manometers | | Volume $cm^3$ | T °C | Q $cm^3$/sec |
|---|---|---|---|---|---|
| | $h_1$ cm | $h_2$ cm | | | |
| 112 | 95.2 | 88.6 | 500 | 28 | 4.464 |
| 112 | 95.2 | 88.6 | 500 | 28 | 4.464 |
| 112 | 95.2 | 88.6 | 500 | 28 | 4.464 |
| 111 | 95.2 | 88.6 | 500 | 28 | 4.505 |
| 111 | 95.2 | 88.6 | 500 | 28 | 4.505 |
| 111 | 95.2 | 88.6 | 500 | 28 | 4.505 |
| 110 | 95.2 | 88.6 | 500 | 28 | 4.545 |
| 111 | 95.2 | 88.6 | 500 | 28 | 4.505 |

## 3. Sunjoto well shape factor

Shape factor is a value that serves as diameter of casing, length of perforated casing, base condition of casing (pervious or impervious), the tip of casing position to the layer position of aquifer, and confined or unconfined aquifer [12]. In 1988, Sunjoto developed a formula to calculate the recharge system dimension under unsteady flow conditions. Sunjoto modified this formula and developed shape factor as the parameter of this equation in order to determine hydraulic conductivity of soils in 2002 and 2014 [13].

$$S = \frac{Q}{FK} \tag{1}$$

Where: $K$ = hydraulic conductivity (m/s), $Q$ = discharge ($m^3$/s), $F$ = shape factor (m), $s$ = drawdown (m). The $F$ parameter of Sunjoto equations represents the well conditions of geometry. Once the research has been conducted, this equation proves that drawdown and hydraulic conductivity value can be computed without graphic and complicated method. Sunjoto continuously developed the $F$ well shape factor under various kinds of well condition. The shape factor $F$ for unconfined aquifer is;

$$F = \frac{2\pi L}{Ln\{\left(\frac{L+2r}{r}\right)+\sqrt{\left(\frac{L}{r}\right)^2+1}\}} \tag{2}$$

Where: $L$ = Length of the Screen (m), $r$ = Radius of Well (m), $F$ = shape factor (m). As it is known as a new formula, this shape factor equation needs to be proved using appropriate well conditions. In this research, the researcher used this equation under unconfined aquifer with pump test.

## 4. Data analysis by shape factor

The pump recharge (full penetration) test (1) has been conducted on May (10), 2019 for (8) hours, the test (2) has been conducted on May (12), 2019 for (12) hours, and test (3) on May (14), 2019 for more than (14) hours at Hydraulic Lab, Department of Civil and Environmental Engineering, UGM. The discharge rate $Q$ is 0.000389105 $m^3$/minutes for each test. In this test, the initial thickness is 0.608 for test (1), 0.359 for test (2), and 0.355 for test (3). The screen length also 0.3 m for test (1) and 0.15 m for test (2) and test (3) (Table.2). According to the Sunjoto method, shape factor value has to be used to determine hydraulic conductivity value. The well geometry is important in order to have accurate calculation. The Hydraulic Conductivity value can be determined using iteration method in Sunjoto shape factor equation. Total value of the last ten values were taken and served as the average value and determine the $K$ value. According to this calculation, the $K$ value is 0.000131 m/sec in test (1), 0.0001809 m/sec in test (2) and 0.000119 for test (3) respectively.

## 5. Results and Conclusion

In this research, the researcher has performed comparison to the hydraulic conductivity value using constant head test and Sunjoto method. The comparison result is acceptable despite of a little difference between disturbed sample and un-disturbed sample (Table.2 and Figure.2). The shape factors in Sunjoto equation have to be used correctly in this analysis. Many researchers developed the shape factor equations on different well conditions. The shape factor equations for unconfined aquifer have been used in this research, and Sunjoto developed this equation under full penetration test in unconfined aquifer. The calculation of shape factor value shows two different results that based on well geometry (length and radius of well). The K value has the same result under pump test despite of the use of different shape factor values. There have some types of error as conditional error, computational error and the assumptions of formula error even though the result can be assumed Sunjoto shape factor equation can be used to determine hydraulic conductivity value under pump test. According to these reasons, the researcher will try to do data analysis using Sunjoto shape factor equation under slug test in the future as described in the introduction part.

**Table 2.** Comparison of hydraulic conductivity value by constant head test and Sunjoto equation

| No | Pump Recharging Test | Q ( Discharge Rate) (m3/min) | Screen length (m) | Initial Thickness (m) | Hydraulic Conductivity Value ( m/sec) | |
|---|---|---|---|---|---|---|
| | | | | | *Constant Head Test* | *Sunjoto Method* |
| 1 | 10-May | 0.000389105 | 0.3 | 0.608 | **0.00038** | **0.000131434** |
| 2 | 12-May | 0.000389105 | 0.15 | 0.359 | **0.00038** | **0.0001801** |
| 3 | 14-May | 0.000389105 | 0.15 | 0.355 | **0.00038** | **0.00011966** |

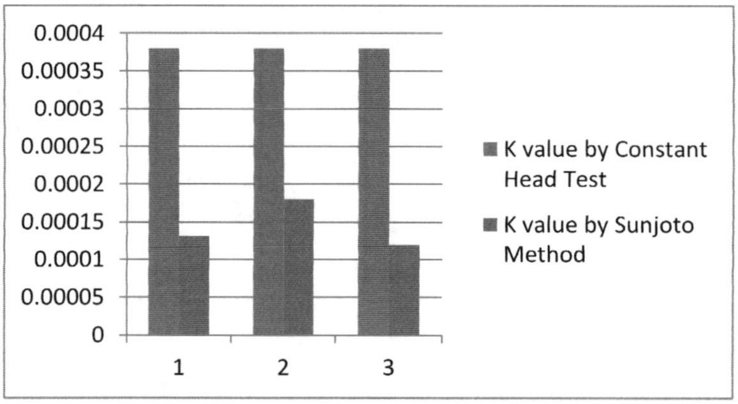


**Figure 2.** Comparison Graph of $K$ value by Constant head test and Sunjoto Method

**Acknowledgments**
The authors wish to thank the supervisor of this research, Professor Sunjoto, Professor Djoko Luknanto from Universitas Gadjah Mada and Professor Yasuto Tachikawa from Kyoto University. This work was partly supported by AUN/SEED-Net Doctoral Degree Sandwich Program (JICA) grant.

**References**

[1] Hvorslev M..J. *Time Lag and Soil Permeability in Ground-Water Observations*, No. 36. Waterways Experiment Station, Corps of Engineers, US Army, 1951.
[2] Brand E.W, and Premchitt J. "Shape factors of cylindrical piezometers", *Geotechnique* 30, no. 4, pp. 369-384, 1980.
[3] Chapuis R.P. "Shape factors for permeability tests in boreholes and piezometers", *Groundwater* 27, no. 5, pp. 647-654, 1989.
[4] Sunjoto S. Influence of Shape Factor to the Hydraulic Pumping Power, In: *20th Congress of the IAHR APD 2016*, Colombo Sri Lanka 28 August-31 September. 2016.
[5] Sunjoto S. Recharge Wells as Drainage System to Increase Groundwater Storage, In: *13rd IAHR-APD Congress, Advance in Hydraulics Water Engineering, Singapore*, 6-8 August 2002. 2002. p. Vol. I, pp.511–4.
[6] Sunjoto S. "The inventions technology on water resources to support environmental engineering based infrastructure", In: *AIP Conference Proceedings*, vol. 1818, no. 1, p. 020055. AIP Publishing, 2017.
[7] Luthfiana A.B. *Determination of in situ soils permeability based on shape factor development*, Bachelor Thesis, UGM, 2015.
[8] Permana A.A. *Determination of Soil permeability in unsatureated condition based on constant discharge by shape factors developmen*t, Bachelor Thesis, UGM, 2015.
[9] Rantu Kexia G.A., Sunjoto S., Hendrayana H. Pumping Test Analysis by Cooper-Jacob and Sunjoto method, *PIT-HATHI XXXIII- Semarang*, 25-27 November, 2016.
[10] Todd D. *Groundwater Hydrology*, John Wiley Son, New Work; 1976. 304 p.
[11] Stibinger J. *Examples of Determining the Hydraulic Conductivity of Soils, Theory and Applications of Selected Basic Methods*. 2014;72.
[12] Sunjoto S. Pengembangan Model Hidraulik Aliran Bawah Permukaan, Laporan Penelitian *PAU_IT-UGM-1988/1989*. 1989.
[13] Sunjoto S. *Groundwater Engineering*, Yogyakarta, Indonesia; 2016. 203 p.

IOP Conf. Series: Materials Science and Engineering **897** (2020) 012013 doi:10.1088/1757-899X/897/1/012013

# Assessment of tool platform micro vibrations induced by moving vehicles in hi-tech factories

**C L Lee**

School of Civil Engineering and Architecture, Xiamen University of Technology, No. 600, Ligong Road, Jimei District, Xiamen, China

Email: cllee63@hotmail.com

**Abstract.** This study explores the micro vibrations of floors and tool platforms induced by internal moving automated guided vehicles (AGVs) in hi-tech factories. The equation of motion of a simplified multi-span floor (or beam) system installed with a tool platform under AGV moving forces generated by a modified Kanai-Tajimi power spectral density (MKT-PSD) function is derived. Dynamic time history analyses of the continuous beam model travelled by different AGV weights are performed. The corresponding root-mean-square (RMS) floor and platform vibration spectra are obtained by using a one-third octave band spectrum analysis and in turn used to compare with the micro vibration criterion. Simulated results indicated that the floor and platform vibrations increase with the AGV weight. For the maximum AGV weight considered, the floor vibration exceeds VC-A, while under the minimum AGV weight, the floor vibration reaches VC-B which is required by the vibration-sensitive tools. Moreover, the platform vibrations are far more than VC-A regardless of the AGV weight. Therefore, the introduction of a proper vibration control scheme is recommended to suppress the excessive micro vibrations of the floors and taller tool platforms.

## 1. Introduction

The fabrication process of silicon wafers and glass panels (or liquid crystal displays, LCDs) with micro or nanometer feature size is susceptible to micro vibrations induced by a variety of disturbance sources. Vibrations directly caused by automated guided vehicles (AGVs) [1] or moving cranes [2] on the floors in hi-tech factories have rarely been studied when designing facilities. Consequently, most of the LCD factories encounter AGV-induced floor and platform vibrations that exceed the desired micro vibration criterion [3]. The LCD factory with thicker RC "cheese" slabs supported by a long span mega truss considered in this study is illustrated in Figure 1. A simplified continuous floor (or beam) model consisting three spans installed with a tool (or equipment) platform at the mid span will be used to quickly simulate the floor-platform system [1].

This study intends to assess the floor and platform micro vibrations caused by internal AGVs in hi-tech LCD factories. The AGV moving force time histories of five different AGV weights with a speed of 2.0 m/s were simulated by using a modified Kanai-Tajimi power spectral density (MKT-PSD) function [4]. Dynamic time history analyses of a simplified three span continuous floor system installed with a platform under AGV moving forces were first carried out by using the state space procedure (SSP) [5]. The corresponding root-mean-square (RMS) micro vibration spectra of the floor and platform were then obtained by performing a one-third octave band analysis [4]. Finally, the

IOP Conf. Series: Materials Science and Engineering **897** (2020) 012013 doi:10.1088/1757-899X/897/1/012013

vibration levels of floor and platform under different AGV moving conditions are assessed and discussed.

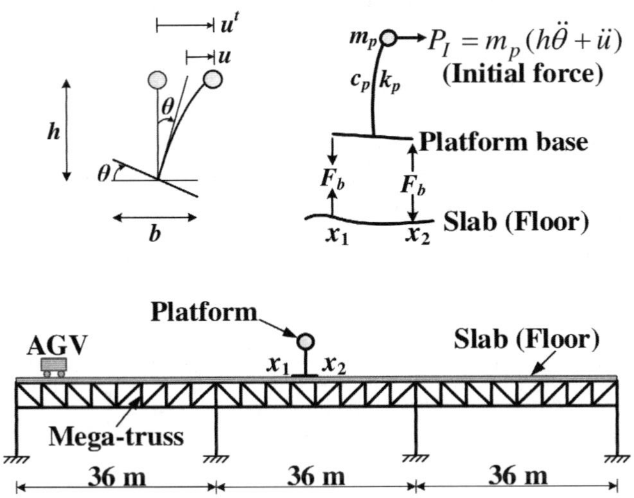


**Figure 1.** Floor system of an LCD factory and numerical model of the platform.

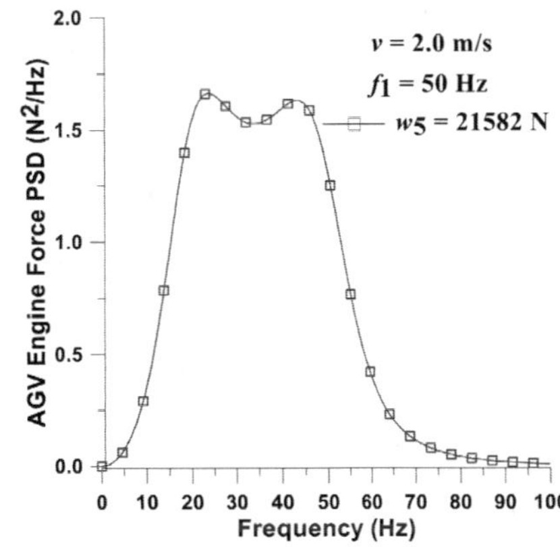


**Figure 2.** Dynamic engine force spectrum.

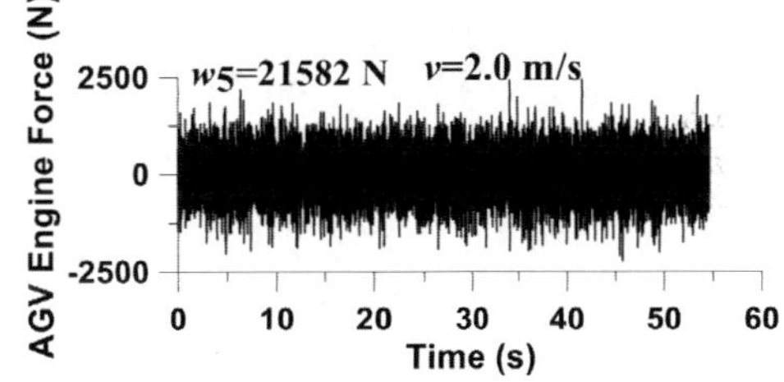


**Figure 3.** Dynamic engine force time history.

## 2. Numerical model of the floor system and platform

The vertical displacement, $w(x,t)$, of a continuous beam (or floor) system that considers the lowest $n$ vibrational modes can be calculated by using the classical mode superposition method [6] as follows:

$$w(x,t) = \sum_{i=1}^{n} q_i(x)\phi_i(x) \tag{1}$$

in which $q_i(t)$ is the $i$-th modal coordinate and $\phi_i(x)$ denotes the corresponding $i$-th mode shape function [7]. Moreover, the equation of motion of the multi-span continuous beam travelled by $N$ moving axle loads can be represented as [1, 4]:

$$\mathbf{M}\ddot{\mathbf{q}}(t) + \mathbf{C}\dot{\mathbf{q}}(t) + \mathbf{K}\mathbf{q}(t) = \mathbf{E}\mathbf{w}(t) \tag{2}$$

in which $\mathbf{M} = [m_{ij}]$ and $\mathbf{K} = [k_{ij}]$ are respectively the $n \times n$ mass and stiffness system matrices with coefficients calculated as [7]:

$$m_{ij} = \int_0^L \rho A \phi_i(x)\phi_j(x)dx \qquad k_{ij} = \int_0^L EI\phi_i''(x)\phi_j''(x)dx \tag{3}$$

in which $\rho$ is the density, $E$ is the Young's modulus, $A$ is the cross-sectional area, $I$ is the moment of inertia of the cross-section, $\phi_i''(x)$ represents the curvature function of the beam and $\mathbf{C}$ is the $n \times n$ damping matrix. Furthermore, $\mathbf{q}(t)=\left[q_1(t) \quad q_2(t) \quad \cdots \quad q_n(t)\right]^{\mathrm{T}}$ is the $n \times 1$ modal displacement vector, $\mathbf{w}(t)=\left[F_1(t) \quad F_2(t) \quad \cdots \quad F_N(t)\right]^{\mathrm{T}}$ is the $N \times 1$ moving load vector, and

$$\mathbf{E}=\begin{bmatrix} \phi_1[x_{F1}(t)] & \phi_1[x_{F2}(t)] & \cdots & \phi_1[x_{FN}(t)] \\ \phi_2[x_{F1}(t)] & \phi_2[x_{F2}(t)] & \cdots & \phi_2[x_{FN}(t)] \\ \vdots & \vdots & \vdots & \vdots \\ \phi_n[x_{F1}(t)] & \phi_n[x_{F2}(t)] & \cdots & \phi_n[x_{FN}(t)] \end{bmatrix} \tag{4}$$

is the $n \times N$ influence matrix of moving axle loads that vary with the locations of the vehicles.

If a platform with the width of $b$ and height of $h$ is mounted on the floor of the factory within the locations of $x_1$ and $x_2$ (Figure 1), its equation of motion along the horizontal direction under the rotational base excitation ($\theta(t)$) can be represented as [6]:

$$m_p\ddot{u}(t)+c_p\dot{u}(t)+k_pu(t)=-m_ph\ddot{\theta}(t) \tag{5}$$

in which $m_p$, $c_p$ and $k_p$ are respectively the mass, damping and stiffness parameters of the platform simulated as a single-degree-of-freedom (SDOF) system in this study. The angular base acceleration is further represented by using the relative vertical floor acceleration located at $x_1$ and $x_2$ as:

$$\ddot{\theta}(t)=[\ddot{w}(x_2,t)-\ddot{w}(x_1,t)]/b=(\mathbf{\Phi}_1^{\mathrm{T}}-\mathbf{\Phi}_2^{\mathrm{T}})\ddot{\mathbf{q}}(t)/b \tag{6}$$

in which $\mathbf{\Phi}_1^{\mathrm{T}}=\left[\phi_1(x_1) \quad \phi_2(x_1) \quad \cdots \quad \phi_n(x_1)\right]$ and $\mathbf{\Phi}_2^{\mathrm{T}}=\left[\phi_1(x_2) \quad \phi_2(x_2) \quad \cdots \quad \phi_n(x_2)\right]$ are the row vectors that contain the mode shape values of each mode at $x_1$ and $x_2$ respectively. Moreover, the resisting coupling force ($F_b(t)$) between the floor and the base of platform induced by the initial force of the platform can be derived with equation (6) as:

$$F_b(t)=m_p\ddot{u}^t(t)h/b=m_p[h\ddot{\theta}(t)+\ddot{u}(t)]h/b=\begin{bmatrix} \dfrac{m_ph^2}{b^2}(\mathbf{\Phi}_2^{\mathrm{T}}-\mathbf{\Phi}_1^{\mathrm{T}}) & \dfrac{m_ph}{b} \end{bmatrix}\begin{bmatrix} \ddot{\mathbf{q}}(t) \\ \ddot{u}(t) \end{bmatrix} \tag{7}$$

The modal coupling force ($F_b^*(t)$) exerted on the floor system at $x_1$ and $x_2$ can be represented as:

$$F_b^*(t)=(-\mathbf{\Phi}_1+\mathbf{\Phi}_2)F_b(t)=\left\{(\mathbf{\Phi}_2-\mathbf{\Phi}_1)\begin{bmatrix} \dfrac{m_ph^2}{b^2}(\mathbf{\Phi}_2^{\mathrm{T}}-\mathbf{\Phi}_1^{\mathrm{T}}) & \dfrac{m_ph}{b} \end{bmatrix}\right\}\begin{bmatrix} \ddot{\mathbf{q}}(t) \\ \ddot{u}(t) \end{bmatrix}=\left\{\bar{\mathbf{M}}\right\}\begin{bmatrix} \ddot{\mathbf{q}}(t) \\ \ddot{u}(t) \end{bmatrix} \tag{8}$$

The equation of motion of the floor equipped with a platform under the AGV moving loads can be further represented by using equations (2), (5), (6) and (8) as:

$$\left(\begin{bmatrix} \mathbf{M} & \mathbf{0} \\ \mathbf{0} & m_p \end{bmatrix}-\begin{bmatrix} \bar{\mathbf{M}} & \\ -m_p\dfrac{h}{b}(\mathbf{\Phi}_2^{\mathrm{T}}-\mathbf{\Phi}_1^{\mathrm{T}}) & 0 \end{bmatrix}\right)\begin{bmatrix} \ddot{\mathbf{q}}(t) \\ \ddot{u}(t) \end{bmatrix}+\begin{bmatrix} \mathbf{C} & \mathbf{0} \\ \mathbf{0} & c_p \end{bmatrix}\begin{bmatrix} \dot{\mathbf{q}}(t) \\ \dot{u}(t) \end{bmatrix}+\begin{bmatrix} \mathbf{K} & \mathbf{0} \\ \mathbf{0} & k_p \end{bmatrix}\begin{bmatrix} \mathbf{q}(t) \\ u(t) \end{bmatrix}=\begin{bmatrix} \mathbf{E} \\ \mathbf{0} \end{bmatrix}\mathbf{w}(t) \tag{9}$$

The numerical solution for equation (9) will be recursively obtained at each time step by using a state space integration procedure [5].

## 3. Simulation results

The system parameters of a real LCD factory adopted in this study are $\rho A = 9.1505 \times 10^3$ kg/m and $EI = 2.4608 \times 10^{11}$ N-m$^2$ which result in a vertical fundamental frequency of 6.285 Hz [8]. All the equivalent damping ratios of the lowest 12 modes of the floor system are assumed to be 5%. Moreover, the system parameters of the platform mounted at $x$=54 m with a mass of 2500 kg, the natural vibration frequency of 25 Hz and the equivalent damping ratio of 2% are given in this numerical example. The major dimensions of the platform are $b$=2.0 m and $h$=2.6 m.

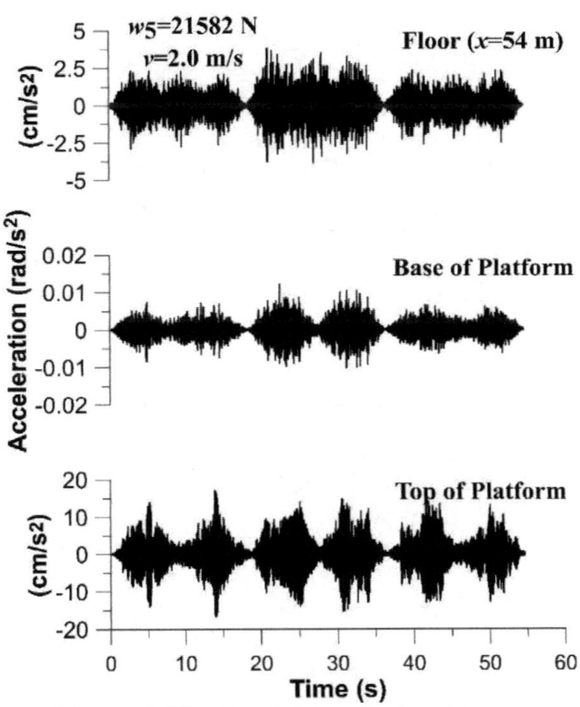

**Figure 4.** Simulated dynamic time history responses of floor system and platform.

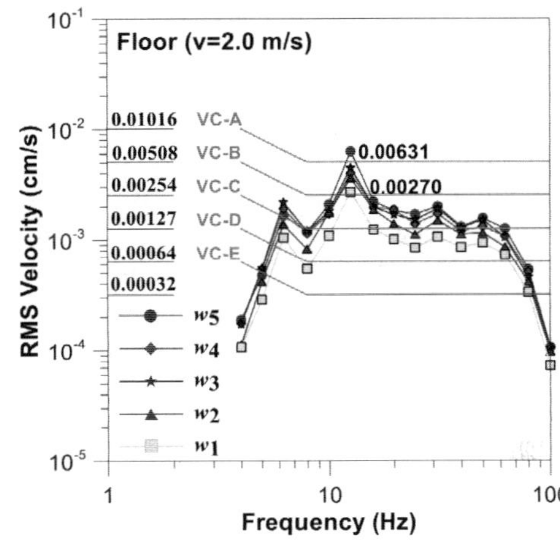

**Figure 5.** RMS floor vibration spectra.

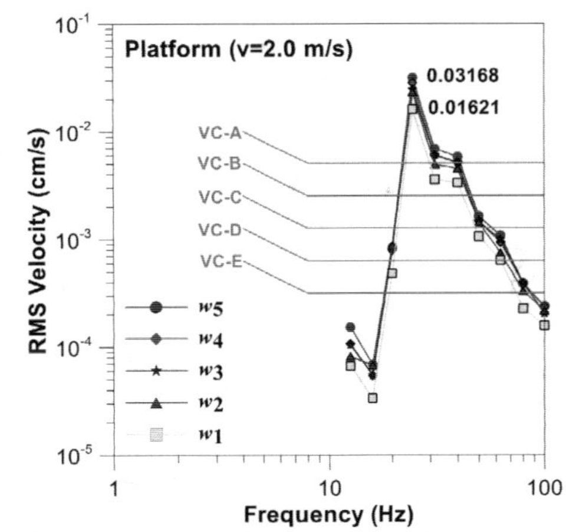

**Figure 6.** RMS platform vibration spectra.

In order to perform the assessment of the floor and platform micro vibrations under a single AGV ($N$=2, i.e. two axle loads) with different weights, five AGV weights with $w_5 = 1.0W =$ 21582 N, $w_4 = 0.8W$, $w_3 = 0.6W$, $w_2 = 0.4W$ and $w_1 = 0.2W$ are considered in this study. Due to the length limitation of the paper, only the simulated dynamic engine force (moving force) MKT-PSD spectrum and its corresponding dynamic engine force time history for an upper bound frequency of $f_1$=50 Hz and the maximum AGV weight ($w_5$) with $v$=2.0 m/s are shown in Figures 2 and 3 respectively [4].

Figure 4 shows the simulated vertical acceleration of the floor system (at $x$=54 m), the angular base acceleration of the platform and the absolute horizontal acceleration of the platform under the maximum AGV weight ($w_5$). The corresponding RMS floor and platform vibration spectra under different AGV weights are shown in Figures 5 and 6 respectively. The floor vibrations, in general, increase with the weight of the AGV. For the maximum AGV weight ( $w_5$ ), the floor vibration exceeds VC-A, while under the minimum AGV weight ( $w_1$ ), the floor vibration reaches VC-B. Moreover, the platform vibrations are far more than VC-A regardless of the AGV weight, and the vibration amplification ratios defined as the ratio of the maximum RMS platform vibration to the maximum RMS floor vibration may reach as high as 5.0 to 7.0 (at around 25 Hz). Therefore, the vibrations of the floors and taller tool platforms adopted by the LCD factory for the large-sized glass panels usually exceed the desired vibration level (VC-B) and the vibration isolation or control devices (dampers) may be introduced to suppress the excessive micro vibrations.

## 4. Concluding remarks

This study has explored AGV induced micro vibrations of tool platforms installed on a three span continuous beam model simplified from a real LCD factory. Time history analyses of the beam model subjected to a pair of moving forces simulated by the MKT-PSD function considering five different AGV weights with $v$=2.0 m/s are performed. The corresponding RMS floor and platform vibration spectra are obtained by using a one-third octave band analysis and further used to compare with the micro vibration criterion. Numerical simulations indicated that the floor and platform vibrations increase with the AGV weight. The floor vibration exceeds VC-A under the maximum AGV weight, while it reaches VC-B (desired vibration level) under the minimum AGV weight. Moreover, the platform vibrations are far more than VC-A regardless of the AGV weight. Therefore, the feasibility of using the vibration control schemes to suppress the excessive micro vibrations of the floors and taller platforms under different AGV weights and vehicle speeds may be evaluated in the future study.

## References

[1] Lee C L, Wang Y P and Su R K L 2013 Assessment of vibrations induced in factories by automated guided vehicles *Proc. Inst. Civil Eng.-Struct. Build.* **166** 182-196
[2] Ju S H, Kuo H H, Yu S W and Ni S H 2019 Investigation of vibrations induced by moving cranes in high-tech factories *J. Low Freq. Noise V. A.* (in press)
[3] Gordon C G 1992 Generic criteria for vibration sensitive equipment *SPIE proceedings* **1619** 71-85
[4] Lee C L, Su R K L and Wang Y P 2013 AGV-induced floor micro-vibration assessment in LCD factories by using a regressional modified Kanai-Tajimi moving force model *Struct. Eng. Mech.* **45** 543-568
[5] Wang Y P, Liao W H and Lee C L 2001 A state-space approach for dynamic analysis of sliding structures *Eng. Struct.* **23** 790-801
[6] Chopra A K 2001 *Dynamics of Structures: Theory and Applications to Earthquake Engineering* (New Jersey: Prentice Hall)
[7] Zheng D Y, Cheung Y K, Au F T K and Cheng Y S 1998 Vibration of multi-span non-uniform beams under moving loads by using modified beam vibration functions *J. Sound Vib.* **212** 455-467
[8] Lee C L and Wang Y P 2019 Assessment of floor micro-vibrations induced by moving vehicles in high technology factories using a fragility-based method *Proc. of the 5th Int. Conf. on Environment Science and Civil Engineering* (Bristol: IOP Publishing) April 13-14 2019 Nanchang China

IOP Conf. Series: Materials Science and Engineering **897** (2020) 012014 doi:10.1088/1757-899X/897/1/012014

# Improved Rapid Assessment of Earthquake Hazard Safety of Structures via Artificial Neural Networks

*Ehsan* Harirchian[1,*] and *Tom* Lahmer[1]

[1]The Institute of Structural Mechanics, Faculty of Civil Engineering, Bauhaus-Universität Weimar, 99423 Germany

**Abstract.**
The vulnerability of structures mainly depends on the structural resistance system of buildings to earthquake. It is unlikely that all existing buildings can be inspected in detail. Therefore, rapid methods for evaluating buildings have been developed over the last decades. This paper investigates the earthquake susceptibility through the combination of buildings' geometrical attributes that affect the vulnerability of building and can be used to obtain an optimal prediction of the damage state of reinforced concrete (RC) buildings using artificial neural networks (ANNs). In this regard, a multi-layer perceptron (MLP) network has been trained and optimized using a database of 145 damaged buildings from the Haiti earthquake. The results demonstrate the practicability and effectiveness of the selected ANNs approach to classify actual structural damage that can be used as a preliminary assessment procedure to recognize vulnerable buildings.

## 1 Introduction

In developing countries, there is a large amount of vulnerable buildings that are to be required for detail earthquake hazard safety evaluation. For instance, in Istanbul-Turkey as a high seismic area, around 90% of buildings are substandard, which can be generalized into other earthquake-prone regions in Turkey [1]. Therefore, it is essential to assess the seismic risk and vulnerability of buildings in urban areas as a primary parameter of the earthquake disaster management policy. Identifying most vulnerable buildings by visual inspection would reduce the time and cost of detailed evaluation. The term "seismic vulnerability" is defined as "the susceptibility of a population of buildings to undergo damage due to seismic ground motion" [2]. Rapid Visual Screening (RVS) is the simplest method among other vulnerability assessment methods, performing necessary structural estimates for quick evaluation of a large building stock [3]. Nevertheless, more detailed analysis might provide a better assessment, but such an approach would entail complicated challenges when an urban scale mitigation campaign is considered. This study introduces a method for improving rapid vulnerability assessment by considering only the RC buildings. For the applicability of the proposed methodology, Haiti, a Caribbean country, which is in a highly active seismic region and has experienced a severe earthquake recently, has been selected.

*e-mail: ehsan.harirchian@uni-weimar.de

## 2 Rapid Visual Screening methods

RVS has been widely used in seismic countries as a practical and simple tool for evaluating the vulnerability of buildings; therefore, this challenge is in the interest of many researchers and still is under developing and improving. In this manner, some efforts have been made to develop national RVS for U.S.A (FEMA 154)[4], India (IITK-GSDMA)[5], Turkey (EMPI)[6], Italy [7], Japan (JPDPA) [8], Philippine [9]. Some practical comparisons between different national RVS methods have been carried out, and their efficiency and robustness have been discussed [10–12]. Other than national and local RVS methods, different proposed methods are popular to use. For instance, an assessment method proposed by Hassan and Sozen [13] for RC buildings, priority for remedial action is manifested in terms of a Priority Index (PI) acquired by adding wall and column indices. Wall index is obtained by normalizing the total area of shear walls and infill walls with the total floor area of the building. Likewise, the column index is achieved by normalizing the total column area with the total floor area. Thus, the method is primarily based on two parameters, the total wall area and the total column area beside the total floor area. Additionally, there are many other RVS methods developed by using linear regression [14–16], Multi-criteria decision making [17], Fuzzy logic [18–20], change detection via images [21], or some other methods concerning their consideration and experiences on a region or country [22]. Still, each of them is based on the expert's opinion, uncertainties, or assumed the linear relationship between parameters. By implementing neural networks, it might get rid of most of the vagueness and uncertainties because the neural networks can compute any function and relationship between variables.

## 3 Data Collection

In January 2010, a catastrophic earthquake with the magnitude $M_w$=7.0 occurred in Haiti. Many of the buildings were damaged, and many more rendered unstable [23]. In this study, a database of 145 damaged RC buildings suffered from the Haiti earthquake has been used from the open-access data on the Datacenterhub platform [24]. The observed structural damage has been classified into four damage grades, as shown in Table 1. Figure 1 illustrates the distribution of damage of the buildings investigated where most of the buildings have severe damage after the earthquake and no building collapsed. Some of the parameters used in this study which are presented in Table 2, are also the basis of the Priority Index proposed by Hassan [13], and this preference was due to the following grounds:

- Their effect has been examined and calibrated for use with structures that do not meet the code requirements for high earthquake risk areas
- They have been tested in areas with different construction practices, and it makes it applicable for different regions
- The required parameters are easy and visible to collect, which leads to less vagueness for investigators

## 4 Application of Artificial Neural Networks to the data

Artificial Neural Networks (ANNs) use their ability to predict future events using historical data of similar events that happened in the past. Traditional RVS methods use a form completed during an observational study of the building. These methods do not use additional data to predict the building vulnerability in case of earthquake events. Therefore, these methods are subjective to the expert that visit the building and subject to human errors. However,

IOP Conf. Series: Materials Science and Engineering **897** (2020) 012014 doi:10.1088/1757-899X/897/1/012014

Table 1: Damage distribution of the investigated buildings

| Damage grade | Extent of damage to buildings | Description of damages |
|---|---|---|
| 1 | Light | Hairline inclined and flexural cracks were observed in structural elements |
| 2 | Moderate | Wider cracks or spalling of concrete was observed |
| 3 | Severe | At least one element had a structural failure |
| 4 | Collapse | At least one floor slab or part of it lost its elevation |

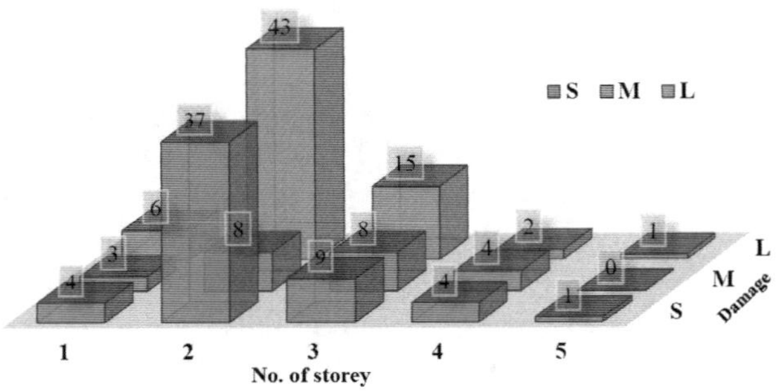


**Figure 1.** Damage distribution of the investigated buildings by the number of storey

Table 2: Independent vulnerability predictors

| Predictor | Unit | Type | Variable |
|---|---|---|---|
| No. of Storey | $N$ | Quantitative | $X_1$ |
| Total Floor Area | $m^2$ | Quantitative | $X_2$ |
| Column Area | $m^2$ | Quantitative | $X_3$ |
| Concrete Wall Area (Y) | $m^2$ | Quantitative | $X_4$ |
| Concrete Wall Area (X) | $m^2$ | Quantitative | $X_5$ |
| Masonry Wall Area (Y) | $m^2$ | Quantitative | $X_6$ |
| Masonry Wall Area (X) | $m^2$ | Quantitative | $X_7$ |
| Captive Columns | $N$ (exist=1, absent=0) | Dummy | $X_8$ |

using ANNs help to improve these methods by considering data of the previous earthquake in the prediction of building states after any earthquake events. The advantage of these methods is that they are swift and reliable if existing data be sufficient for training an ANNs. A multi-layer feedforward neural network, which is called a multi-layer perceptron (MLP), is used to predict the building state after an earthquake. The network was trained by 115 buildings using a k-fold cross-validation method, and the rest 30 of buildings remained for the test. Three-layered (input layer, hidden layer, and output layer) feedforward multi-layer perceptron (MLP) was used and trained with the error backpropagation algorithm. The MLP architecture presented in Figure 2. The implementation of the model is carried out using Scikit-learn as it is the well-maintained, comprehensive, and open-sourced machine learning package in Python programming language.

IOP Conf. Series: Materials Science and Engineering **897** (2020) 012014 doi:10.1088/1757-899X/897/1/012014

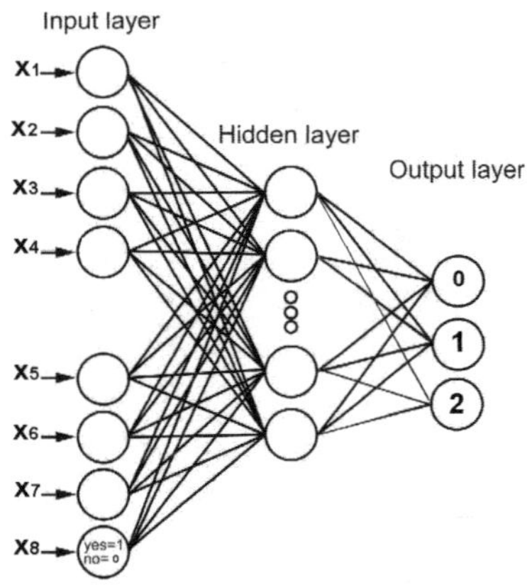


**Figure 2.** Multi-layer perceptron neural network

## 5 Results and discussion

K-fold cross-validation (K=10) was performed to confirm the accuracy of the classification procedure during training procedures. A test dataset of 20% (30 buildings) was held to estimate the generalization capability of the proposed network. The hyper-parameter values of MLP model have been shown in Table 3. A 3-layer MLP (1 Hidden layer) trained with Adam solver and adaptive learning rate with 2000 iteration and 0.01 alpha. Two multi-class metrics have considered to evaluate model performance as a dataset is a multi-class dataset. These are balanced-accuracy (with class-balanced sample weights) and receiver operating characteristics (ROC). ROC curves typically feature a true positive rate on the Y-axis and false positive rate on the X-axis. It means that the top left corner of the plot is the "ideal" point, which means that a larger area under the curve (AUC) is usually better. Table 4 demonstrates the accuracy of the model for training and test dataset. Table 4 illustrates that the model reached an accuracy of 0.56 (or 56%) on the test dataset. Figures 3 to 5 show more details for each class based on the confusion matrix and also the ROC curves of training and test samples. From Figure 4, it is clear that the model can fit the trained data. However, as can be seen in Figure 5 for test data, the model has poor performance in class 1, which means that it cannot handle this class properly in compare to other classes. As it has been observed from the confusion matrix, most of class 0 categorized as class 2, which means that the feature of the building of light class is near to the feature of building in severe class. However, the model is fine for class 1 and class 2, where most of the buildings correctly classified. Figure 5 shows ROC of test dataset. According to this figure, class 1 has AUC of 0.72, which is the lowest AUC within all classes. Figure 5 also illustrates that the micro-average AUC is 0.77, which is a good AUC for such a small dataset. According to Table 4 and Figures 3, 4 and 5 it can be concluded that the model have high ability of classifying light and collapse buildings. Nevertheless, it suffers from classifying moderate building, which is not in high priority in case of rapid assessment where the most important buildings are buildings subject to severe damage or collapse during an earthquake.

Table 3: Hyper-parameter Values

| **Hidden Layer sizes** | **Activation Function** | **Solvers** | **Alpha** | **Maximum Iteration** | **Learning Rate** |
|---|---|---|---|---|---|
| 10 | Relu | Adam | 0.01 | 2000 | Adaptive |

IOP Conf. Series: Materials Science and Engineering **897** (2020) 012014 doi:10.1088/1757-899X/897/1/012014

Table 4: Accuracy of the model on train and test dataset

| Accuracy of train dataset | Accuracy of test dataset |
|---|---|
| 0.76 | 0.55 |

| | Class 0 | Class 1 | Class 2 |
|---|---|---|---|
| Class 0 | 6 | 2 | 5 |
| Class 1 | 1 | 4 | 1 |
| Class 2 | 3 | 1 | 7 |

**Figure 3.** Confusion matrix of the test data (class 0: Light, 1: Moderate, and 2: Severe)

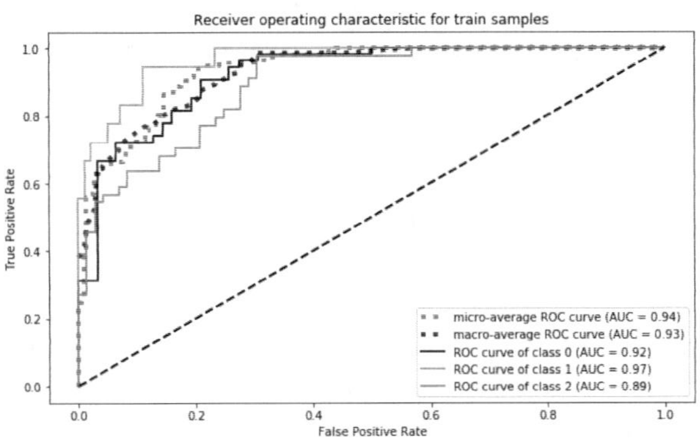


**Figure 4.** ROC of training dataset for all classes (class 0: Light, 1: Moderate, and 2: Severe)

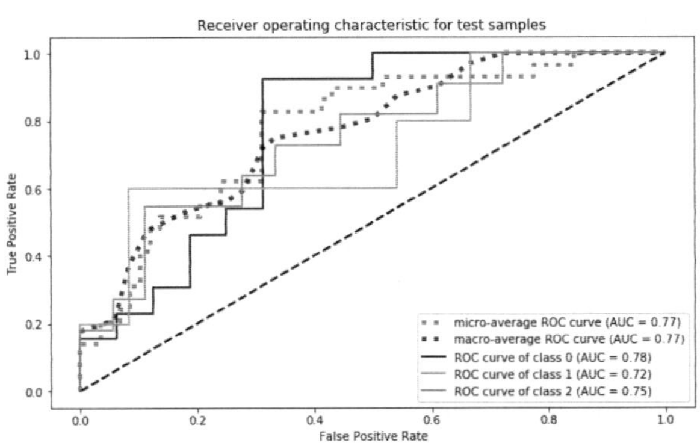


**Figure 5.** : ROC of test dataset for all classes (class 0: Light, 1: Moderate, and 2: Severe)

## 6 Conclusion

In this paper, a practical framework for the Improvement of Rapid Assessment of Earthquake Hazard Safety of structures via Artificial Neural Networks termed shortly IRAEHSAN within the application of eight performance modifiers, has been demonstrated. It must be mentioned that the accuracy of the study depends on the sample buildings chosen, the calculation methods, and the parameters that are present during the ANNs instruction and testing processes; however, in future studies, more data can be used. The method shows its robust ability to classify buildings into light and severe. However, it is weak to classify buildings accurately into moderate damage, which is not in high priority in the case of rapid assessment, where the most important buildings are buildings subject to severe damage or collapse during an earthquake. The proposed method helps to manage and implement strategies for the safety of the communities before an earthquake strike by investigating the vulnerability classes for each building and identifying highly vulnerable buildings that deserve further inquiry.

## References

[1] H. Sucuoğlu, U. Yazgan, A. Yakut, Earthquake Spectra **23**, 441 (2007), https://doi.org/10.1193/1.2720931

[2] A. Shahriar, M. Modirzadeh, R. Sadiq, S. Tesfamariam, Earthquake and Structures **3**, 695 (2012)

[3] P. Kapetana, S. Dritsos, WIT Transactions on the Built Environment **93**, 409 (2007)

[4] FEMA P-154, Applied Technological Council (ATC) (2015)

[5] D.C. Rai, p. 32 (2005)

[6] A. Ansal, K. Özaydın, A. Edinçliler, A. Sağlamer, H. Sucuoğlu, P. Özdemir, Metropolital Municipality of Istanbul, Planning and Construction Directorate, Geotechnical and Earthquake Investigation Department, Turkey (2003)

[7] P. Angeletti, A. Bellina, E. Guagenti, A. Moretti, V. Petrini, *Comparison between vulnerability assessment and damage index, some results*, in *Proceedings of the 9th World Conference on Earthquake Engineering* (1988), Vol. 7, pp. 181–186

[8] S. Evaluation, T.J.B.D.P.A. Retrofit (2001)

[9] C.B. Vallejo et al., *Rapid Visual Screening of Buildings in the City of Manila, Philippines*, in *5th Civil Engineering Conference in the Asian Region and Australasian Structural Engineering Conference 2010, The* (Engineers Australia, 2010), p. 513

[10] E. Harirchian, T. Lahmer, S. Buddhiraju, K. Mohammad, A. Mosavi, Buildings **10**, 51 (2020)

[11] E. Harirchian, T. Lahmer, IOP Conference Series: Materials Science and Engineering **652**, 012069 (2019)

[12] E. Işık, Bitlis Eren University Journal of Science and Technology **5**, 71 (2015)

[13] A.F. Hassan, M.A. Sozen, ACI Structural Journal **94**, 31 (1997)

[14] N. Özhendekci, D. Özhendekci, Earthquake Spectra **28**, 1165 (2012)

[15] S.K. Jain, K. Mitra, M. Kumar, M. Shah, Earthquake Spectra **26**, 709 (2010)

[16] M. Ningthoujam, R.P. Nanda, International Journal of Disaster Risk Reduction **28**, 720 (2018)

[17] E. Harirchian, A. Harirchian, *Earthquake Hazard Safety Assessment of Buildings via Smartphone App: An Introduction to the Prototype Features- 30. Forum Bauinformatik: von jungen Forschenden für junge Forschende: September 2018, Informatik im Bauwesen* (Professur Informatik im Bauwesen, Bauhaus-Universität Weimar, 2018), https://e-pub.uni-weimar.de/opus4/frontdoor/index/index/docId/3785

[18] S. Tesfamariam, M. Saatcioglu, Earthquake Spectra **26**, 235 (2010)

[19] Z. Şen, Expert Systems with Applications **37**, 5653 (2010)

[20] E. Harirchian, T. Lahmer, Applied Sciences **10**, 2375 (2020)

[21] A. CEROVECKI, S. GHARAHJEH, E. HARIRCHIAN, D. ILIN, K. OKHOTNIKOVA, J. KERSTEN, Preface 2 Summer Course 2015 p. 20 (2015)

[22] R.P. Nanda, D.R. Majhi, Journal of The Institution of Engineers (India): Series A **94**, 187 (2013)

[23] H.K. Miyamoto, A. Gilani, *Urban earthquake damage assessment and reconstruction: 2010 Haiti earthquake*, in *Proc., World Conf. on Earthquake Engineering, International Association for Earthquake Engineering (IAEE)* (2012)

[24] C.A. Christine, H. Chandima, P. Santiago, L. Lucas, S. Chungwook, P. Aishwarya, B. Andres, Computing in Science & Engineering, **20**, 49 (2018)

IOP Conf. Series: Materials Science and Engineering **897** (2020) 012015 doi:10.1088/1757-899X/897/1/012015

# Evaluating critical criteria for green hospital buildings

**Hsiang-Hsun Huang [1], Yu-Ying Huang [2*] and Yeng-Horng Perng [1]**
[1] Department of Architecture, National Taiwan University of Science and Technology, No.43, Sec. 4, Keelung Rd., Da'an Dist., Taipei 10607, Taiwan
[2] Department of Real Estate and Built Environment, National Taipei University, No.151,
University Rd., San Shia District, New Taipei City, 23741 Taiwan
[*] Email: D98130O2@gmail.com

**Abstract.** Greening building is an increasingly important concern and a challenge in construction industry. The purpose of this paper is to evaluate the weights of criteria for green hospital buildings that provides decision-makers the priorities strategy in greening process. AHP was applied in this study to establish the hierarchy structure and evaluate the weight of criteria for green hospital buildings. The result indicated that sustainable site, energy saving, and water efficiency are the top three criteria. In the overall ranking, the top three significant sub-criteria are design of layout, greening of site, and energy saving for façade. The contribution of this study is to help decision-makers understand the relatively important of criteria to make decision for green hospital projects for sustainable development.

## 1. Introduction

Hospital is one of the public buildings, which has characteristics of high energy consumption, air and waste pollution. For instance, the air conditioning consumption of hospital is the highest of energy consumption, which accounts over 50%, followed by lighting [1]. Against the background of sustainable development, the strategy of green building practices could solve energy and environmental problems in hospitals. Thus, the design and construction of hospital towards green building that has become an important issue.

Several studies have explored the assessment criteria and design of green hospital. Unger, Campion et al. [2] used checklists to evaluate the quantifiable metrics in green healthcare and indicated that the checklists have helped to advance sustainability in healthcare. Wood, Wang et al. [3] used a questionnaire survey to investigate the users' perceptions of green design elements in hospitals, then they translated the needs to design requirements suitable for implementation. Sahamir and Zakaria [4] investigated green criteria for hospital building development in Malaysia and compared green assessment criteria with different rating systems, there by indicating that energy efficiency is the most important criteria of green hospital building. However, there was rarely research focused on the relative importance of criteria in green hospital building. Consequently, this study applied the Analytic Hierarchy Process (AHP) method to assess the relative importance of criteria for reducing energy consumption and environmental impacts in hospital.

## 2. Hierarchy model

To consider the local situation and regional development of Taiwan, this paper referenced literature review and the green rating system of US Leadership in Energy and Environmental Design (LEED) to establish the hierarchy structure for green hospital buildings, which was suitable for Taiwan. The AHP method was applied in this study to establish the criteria hierarchy. The hierarchy structure consisted main objective, multi-criteria and sub-criteria, as shown in Fig.1. The main objective at first hierarchy level that was green hospital buildings that provided ecology, water-saving, energy-saving, waste reduction and healthy environmental for patient and staff. The multi-criteria at second hierarchy level that contained sustainable site, water efficiency, energy-saving, materials and resources, and indoor environmental quality. Within the 5 criteria, 23 sub-criteria were identified at the third hierarchy level, as shown in Fig1. Therefore, relevant criteria and sub-criteria for green hospital building were identified.

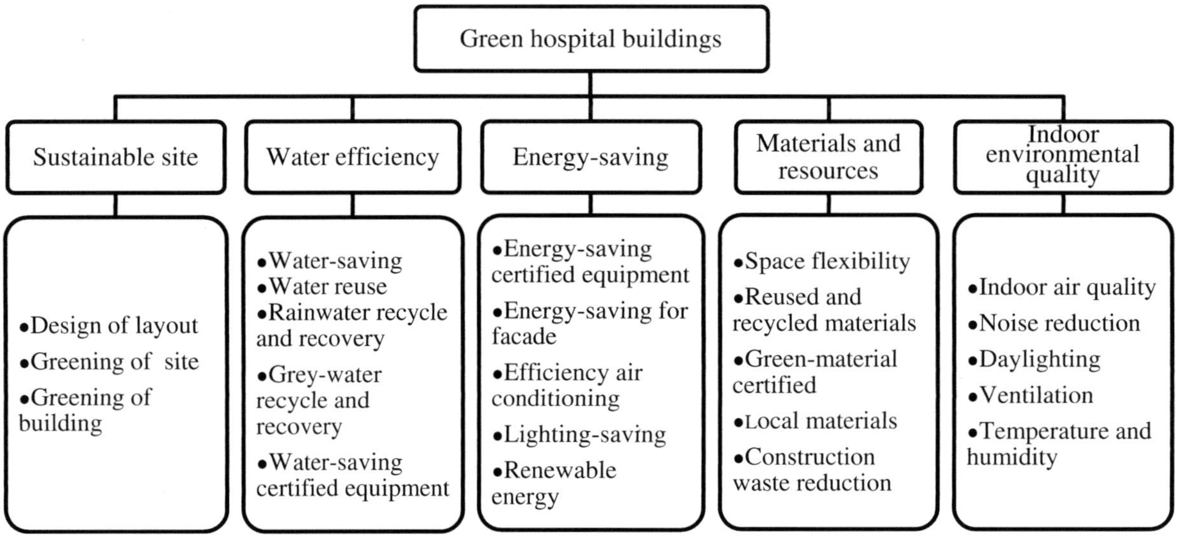


**Figure 1.** The hierarchy structure

## 3. Methodology

AHP is one of the multi-criteria decision making (MCDM) methods to help decision-makers dealing with complex problems with multiple criteria, which can formulate hierarchy structure and measure weight for ranking. Several studies have applied AHP method to evaluate the weight of criteria in different fields. Wannakul, Pathumnakul et al. applied AHP method to evaluate the weight of criteria that affecting decision to choose government R&D projects for small and medium-sized enterprises [5]. Ghimire and Kim used AHP methodology to estimate and rank the barriers to developing renewable energy in Nepal [6]. Wang and Perng used AHP method to identify the key indicators affecting customer satisfaction for the interior design industry [7]. Tse-Hsiung and Perng Perng adopted AHP method to explore the Key factors affecting the rent of office buildings in Taipei city [8]. Lin and Perng used AHP method to explore the key indicators in the space design of commercial [9]. In order to achieve the objective in this study, AHP was applied to evaluate the weight of criteria for reducing the energy consumption and improving environmental for hospital. The AHP proceeded as follows [10][11]:

- Construct a hierarchy structure with main objective, multi-criteria and sub-criteria.
- Assess relative importance of criteria by pairwise comparison and build the judgment matrix.
- Evaluate the local and global weight of criteria and confirm the sum of weight is equal to 1.
- Calculate Consistency Index (CI) and Consistency Ratio (CR) of expert evaluation matrix.
- Verify the value of CI and CR is less than 0.1, the expert matrix is regarded as consistency.
- Ranking the local and global weight of criteria for decision-making.

## 4. Analysis result

The analysis data of pairwise comparison questionnaires were collected from 45 experts in Taiwan. Table 1 summarized the local weight of the criteria and sub-criteria from experts' evaluation. All the judgment matrixes were verified the calculated value of CI and CR that less than 0.1. Rank columns displayed the priority of the criteria and sub-criteria. As shown in Table 1, the top three importance of criteria are sustainable site (0.28), energy saving (0.25), and water efficiency (0.21). By contrast, indoor environmental quality (0.13), materials and resources (0.12) were less importance of criteria, respectively.

According the local weight of sub-criteria, indicating that design of layout (0.49), was found to be the most important sub-criteria within the sustainable site, followed by greening of site (0.31) and greening of building (0.2), respectively. Within the water efficiency, water-saving plan (0.27) was found to be the most important sub-criteria, followed by water reuse (0.26). Within the energy efficiency, energy-saving for façade (0.27) was found to be the most important sub-criteria, followed by efficiency air conditioning (0.26). Meanwhile, Table 1 indicated that space flexibility, construction waste reduction, reused and recycled materials (0.22) were the most important sub-criteria within the materials and resources. Within the indoor environmental quality, indoor air quality (0.27) was the most important sub-criteria, followed by noise reduction and ventilation (0.21).

**Table 1.** Criteria weights.

| Criteria | Local weight | Rank | Sub-criteria | Local weight | Rank |
|---|---|---|---|---|---|
| Sustainable site | 0.28 | 1 | Design of layout | 0.49 | 1 |
| | | | Greening of site | 0.31 | 2 |
| | | | Greening of building | 0.20 | 3 |
| Water efficiency | 0.21 | 3 | Water-saving | 0.27 | 1 |
| | | | Water reuse | 0.26 | 2 |
| | | | Rainwater recycle and recovery | 0.21 | 3 |
| | | | Grey-water recycle and recovery | 0.18 | 4 |
| | | | Water-saving certified equipment | 0.08 | 5 |
| Energy-saving | 0.25 | 2 | Energy-saving certified equipment | 0.11 | 5 |
| | | | Energy-saving for facade | 0.27 | 1 |
| | | | Efficiency air conditioning | 0.26 | 2 |
| | | | Lighting-saving | 0.20 | 3 |
| | | | Renewable energy | 0.17 | 4 |
| Materials and resources | 0.12 | 5 | Space flexibility | 0.22 | 1 |
| | | | Reused and recycled materials | 0.22 | 1 |
| | | | Green-material certified | 0.16 | 5 |
| | | | Use local materials | 0.18 | 4 |
| | | | Construction waste reduction | 0.22 | 1 |
| Indoor environmental quality | 0.13 | 4 | Indoor air quality | 0.27 | 1 |
| | | | Noise reduction | 0.21 | 2 |
| | | | Daylighting | 0.16 | 4 |
| | | | Ventilation | 0.21 | 2 |
| | | | Temperature and humidity | 0.15 | 5 |

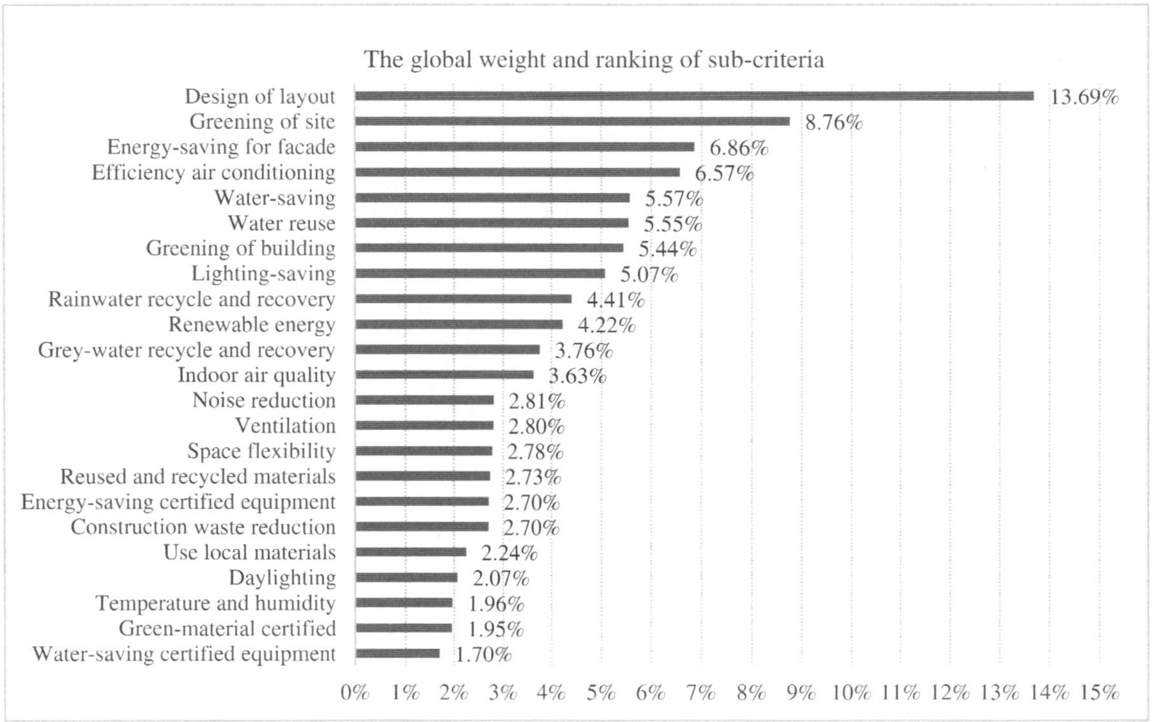


**Figure 2.** Overall ranking of sub-criteria

Moreover, this study presented the global weight and ranking of sub-criteria, as shown in Fig.2. All the comparison of sub-criteria has been performed. The top five significant sub-criteria were design of layout, greening of site, energy-saving for façade, efficiency air conditioning, and water-saving. The figure clearly showed that design of layout was the most significant sub-criterion, which accounted for 13.69% of the total global weight. In the overall ranking, water reuse, greening of building, lighting-saving, rainwater recycle and recovery, and renewable energy were ranked sixth through tenth, respectively. By contrast, the least significant sub-criterion is water-saving certified equipment, which accounted for 1.7% of the total global weight. The result of ranking may help decision-makers to make effective and efficient decisions in limited budget and time for green hospital buildings.

## 5. Conclusion

AHP was applied in this study to establish the criteria hierarchy and the relation importance of all criteria for green hospital buildings. The analysis result in this paper indicated that the top three importance of criteria are "sustainable site", "energy saving", and "water efficiency" for green hospital building evaluation. "Indoor environmental quality", "materials and resources" were less importance of criteria, respectively. According the analysis result of sub-criteria, indicating that "design of layout" was found to be the most important sub-criteria within the sustainable site criteria. "Water-saving plan" was found to be the most important sub-criteria within the water efficiency criteria. "Energy-saving for façade" was found to be the most important sub-criteria within the energy efficiency criteria. "Space flexibility", "construction waste reduction", "reused and recycled materials" were the most important sub-criteria within the materials and resources criteria. "Indoor air quality" was the most important sub-criteria within the indoor environmental quality criteria. In the overall ranking, the most significant criterion has been discovered as "design of layout", followed by "greening of site", "energy saving for façade", and "efficiency air conditioning", respectively. By contrast, the last significant criterion is "water-saving certified equipment". The analysis result of the relatively important and ranking may help decision-makers to make effective and efficient decisions in limited budget and time for green hospital buildings that practical towards sustainable development.

The research limitations in this study include local situation, regional, and the green rating system development in different countries. Therefore, the analysis result in this paper is limited to the green hospital buildings in Taiwan; different countries require individual criteria and analyses for green hospital buildings. Otherwise, there are 23 sub-criteria under 5 main criteria, some more criteria have not been identified and categorized. For future work, the research can be done by using other MCDM methods like Fuzzy AHP, Analytic Network Process (ANP) to dealing with similar or another issue with multiple criteria.

**References**

[1] Ji R and Qu S 2019 Investigation and Evaluation of Energy Consumption Performance for Hospital Buildings in China *Sustainability.* **11(6)** 1724
[2] Unger S R, Campion N, Bilec M M, and Landis A E 2016 Evaluating quantifiable metrics for hospital green checklists *Journal of Cleaner Production.* **127** 134-142
[3] Wood L C, Wang C, Abdul-Rahman H, and Jamal Abdul-Nasir N S 2016 Green hospital design: integrating quality function deployment and end-user demands *Journal of Cleaner Production.* **112** 903-913
[4] Sahamir S R and Zakaria R 2014 Green Assessment Criteria for Public Hospital Building Development in Malaysia *Procedia Environmental Sciences.* **20** 106-115
[5] Wannakul R, Pathumnakul S, and Arunyanart S 2019 Factors Affecting Decision to Select Government R&D Support Projects for SMEs *IOP Conference Series: Materials Science and Engineering.* **639** 012011
[6] Ghimire L P and Kim Y 2018 An analysis on barriers to renewable energy development in the context of Nepal using AHP *Renewable Energy.* **129** 446-456
[7] Wang P-K and Perng Y-H 2019 Critical Criteria for Customer Satisfaction of Interior Design *IOP Conference Series: Materials Science and Engineering.* **652** 012012
[8] Tse-Hsiung L and Perng Y-H 2019 Key factors affecting the rent of office buildings: case study on Taipei 101 *IOP Conference Series: Materials Science and Engineering.* **652** 012066
[9] Lin M-F and Perng Y-H 2019 Exploration of key commercial space design indicators *IOP Conference Series: Materials Science and Engineering.* **652** 012067
[10] Saaty T L 1988 *What is the Analytic Hierarchy Process?* (Berlin, Heidelberg: Springer Berlin Heidelberg) 109-121
[11] Saaty TL 1980 The Analytic Hierarchy Process (New York: McGraw-Hill)

IOP Conf. Series: Materials Science and Engineering **897** (2020) 012016 doi:10.1088/1757-899X/897/1/012016

# Implementation of Design and Build Contract in Government Building Construction Project Practice

**Anggoro Putro[1] and Yusuf Latief[1]**

[1]Civil Engineering Department, Faculty of Engineering, University of Indonesia, Depok, West Java, 16424

Email: jagadnegoro2002@gmail.com

**Abstract.** The Development of government infrastructure projects, which has grown rapidly in recent years, require the Indonesian Government, through the Ministry of Public Work and Housing, to socialize the breakthrough of the national construction procurement system with design and build contract. This breakthrough on construction projects is expected to accelerate the achievement of government infrastructure targets, thus supporting national economic growth. The purpose of this research is to identify DB characteristics, implementation, the advantages and disadvantages in Indonesian construction projects, and to analyze the obstacles, conflicts and problems emerging on the construction project implementation. In the construction project implementation, all participants involved hopes that the project can be completed within the objectives. The expected objectives are that the project is completed in time, does not exceed the budget and meets the quality requirements.

## 1. Introduction

The increasing amount and volume of infrastructure developments accompanied by the demand for faster work practice brings about the need for the modernization of the national construction work procurement system and project delivery that makes it easier for the government to achieve their goals. Development trends of the Ministry of Public Works and Housing will have a significant contribution in the national economic growth. Faster fund absorption and project accomplishments determines the speed of manpower and equipment which will in turn determine the rate of national economic growth.

The design and build contract method is considered to be able to increase the speed of which projects can be accomplished. The fast track system in the design and build contract allows for a trimming of duration in the auction service provider (contractors) which does not need to be carried out separately with the selection of planning consultant services. Construction projects implemented with the design and build contract can at least reduce the time duration for contractor auctions. Other advantages of using the design and build contract include the two tasks lying under one responsible party, project cost efficiency without reducing quality, reduces project duration, increases communication, and reduces lawsuit and litigation related to project implementation since project risks and costs are under the responsibility of the selected contractor [1].

The purpose of this research is to identify design and build characteristics, implementation, the advantages and disadvantages in Indonesian construction projects, and to analyze the obstacles, conflicts and problems emerging on the construction project implementation. According to Azizan and Ibrahim [2], design and build contract is classified as the project with most risks. This is due to the combination of design activities, construction and monitoring process, and the participation of contractors and clients in the construction projects has a high level of risk. The outcome of this paper are expected to be supporting and fundamental data, which will be taken to the next level of main research, namely the development of the Standard Method of Measurement (SMM) based on Work Breakdown Structure which is applied to the integrated design and build contract for state and government building construction projects in increasing accountability of audits of the auditing agency and the internal auditor of the ministry the institution.

## 2. Theoretical framework

### *2.1. Design and build definition*

As mentioned in the Ministry of Public Works and Housing Regulation Number 12 Year 2017, Design and build Construction Works are all works related to the construction of a structure where the design process is integrated with the construction. In other words, this construction work is carried out by more than one contractor. Clause 1 chapter 1 point 13 states that a design and build contract is a written agreement between the owner and the contractor.

Design and build contract can be identified as a procurement method where one entity or consortium is contractually responsible for both the design and the construction of a project [3]. Hale and Shrestha [4] describes the design and build contract as a project delivery method in which the owner provides requirements for the specified project and awards a contract to one company who will both design and build the project [3]. The design and build procurement approach had been identified to be rapidly growing and patronized in the global construction industry. This is due to the several benefits that this procurement approach provides over the other procurement approaches, most notably the traditional procurement option, which is characterized by inherent fragmentation which leads to time and cost overruns. Gambo and Gomez [3] also mentioned that the design and build procurement approach is different from other procurement approaches due to its advantages of offering single point responsibility, inherent build ability, and risk allocation.

Design Build Institute of America (DBIA) states that the design and build method is an implementation method of project delivery where the contractor integrates a single contract with the owner to provide architectural or engineering design and construction services. The design and build contract is also known as the sole responsibility concept. Without the design and build approach, the owner's architect or engineer commissioner prepares drawings and specifications based on the design contract, then selects a contractor with a competitive bid (or negotiation) to build the facility based on user requirements [5]. Figure 1 shows the difference between the design and build method and the conventional method.

Mentioned by Ling and Leong [6] in their journal, several studies regarding design and build project performance have been done. In the UK, a project owner declared that they have obtained above-average satisfaction with a design and build project in terms of cost, time and quality of performance [7]. Another study showed that the design and build project had greater time and cost certainty, better value for money, and 50% more likely to be completed on time and in accordance with the agreed budget compared to the design-bid-build project [8]. A study conducted in the USA by Konchar and Sanvindo [9] stated that time, cost, and quality performance of design and build projects are generally better than design-bid-build projects.

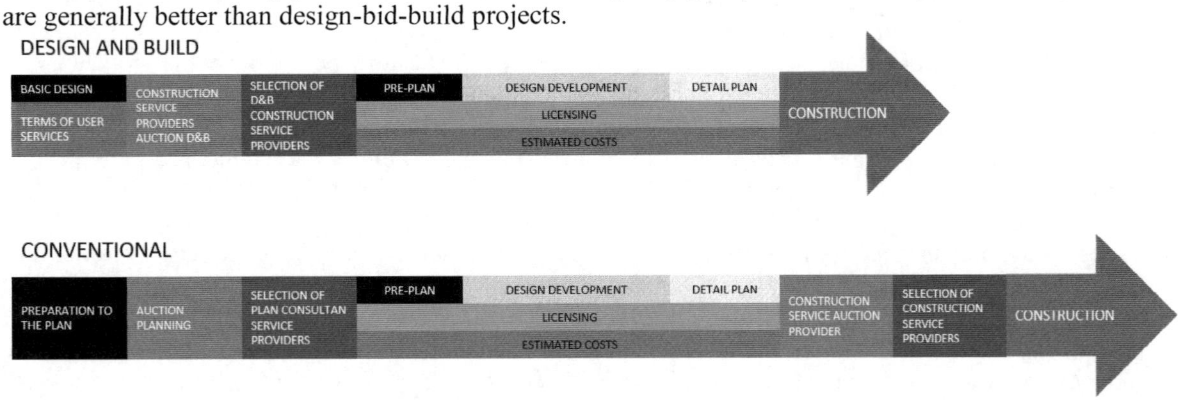


**Figure 1.** The difference in the flow of two methods.

### *2.2. Design and build characteristic*

Contextually, design and build provide a single responsibility point for design and construction service. This is considered correct according to the contractor's perspective, but not from the contracted company, which may subcontract services of which they have no expertise, or join a companies' consortium [10]. Besides a single responsibility point, this method has several other features. Park et.al. [11] mentioned that in a big and complex company this method benefits good coordination and communication.

Gambo and Gomez [3] in their journal defines the design and build characteristic as in the Table 1.

**Table 1.** Design and build characteristic.

| Characteristics of D&B procurement approach | Authors |
|---|---|
| Single Point Responsibility; | Seng & Yousof, (2006); Skitmore & Ng, (2002) |
| Complexity; | Abdulrashid, (2002); Oztas & Okmen, (2004) |
| Risk Allocation; | Hassim et., al. (2008); Muhammad, (2005); Beard et., al. (2001) |
| Compressed Delivery Schedule; | Abdulrashid, (2002); Chan & Yu (2005) |
| Communication; | Chan et., al. (2001); Levy, (2006) |
| Effective Client Representation; | Lam et., al. (2004); Peterson & Murpheree, (2004) |
| Facilitates use of latest innovative technologies. | Ling, Chong & Ee (2004); Abdulrashid, (2002) |

Table 1 shows the characteristics of the design and build contract obtained based on summaries from previous journals and literature in Gambo and Gomez's research in 2015[3]. Design and build approach allow a client to directly contract the contractor without an intermediary consultant, and the contractual position of the project lies between the client and the contractor [12]. A single responsibility point featured in the design and build contract causes contractors to be fully responsible for the performance of completed projects, even if the problems or errors that arise are caused by the activities of subcontractors involved in the development process.

## 3. Findings and discussion

### *3.1. Design and build implementation*

According to Research and Development Agency of Ministry of Public Work and Housing in 2017 Design and build contract is a construction procurement method that combines planning, execution, and monitoring process in a single service provider. Design and build contract's advantages utilize a lump sum payment method to transfer all risks to the service provider and streamlining the process of implementing the procurement of goods or services. Extended warranty period is another advantage of this method. Moreover, a lump sum contract allows a flexibility for volume adjustment because what is specified in the contract document is the design criteria (basic design or concept design).

With the benefits, come the disadvantages. The design and build contract method have several significant disadvantages such as a higher project cost [13], less variation flexibility [14,15], and uneconomic use of resources. Projects are likely to undergo changes due to specification-related issues, long and time-consuming revision process [16], risks due to the contractors' limitations regarding the government's laws and sovereignty, as well as the risk that the government does not abide to the agreed contract—a risk that arises due to disasters, technical issues, and changes of political and economic scenarios [10]. Here in Table 2 is a few examples of DB implementation in Indonesian construction companies that shows that DB contracts do not always guarantee the work will be on time.

IOP Conf. Series: Materials Science and Engineering **897** (2020) 012016 doi:10.1088/1757-899X/897/1/012016

**Table 2.** Design and build implementation in Indonesia.

| Design and build Project | Service Provider | Planned Implementation Period | Notes |
|---|---|---|---|
| Renovation of the Adi Sutjipto Airport Terminal in Yogyakarta | PT. Wijaya Karya | 3 Months | On schedule work |
| Adi Soemarmo Airport Terminal accomplished work | PT. Wijaya Karya | 7 Months | 180 days extended |
| Office Room Regulation | PT. Sarijai Adhitama | 5 Months | 150 days extended |
| Hassanudin Makassar Airport Terminal accomplished work | PT. Adhi Karya jo. PT. Tridaya jo. PT. Dedato | 6 Months | 120 days extended |
| Regulation of Sam Ratulangi Manado Airport Terminal | PT. Sarana Bangun Persada | 6 Months | 60 days extended |

Besides the examples above, there are some disadvantages that can cause problems and reduce contractor accomplishments. As for some examples of problems found by the Indonesian institutional auditor and problems found in previous journals are shown below in Table 3 dan Table 4 consecutive.

**Table 3.** Problems arising from the implementation of design and build in Indonesia.

| Project Title | Type of Contract | Problems |
|---|---|---|
| Bogor Ring Road Highway Project | PT JM standard DB Contract | • There is no clear separator between the terms of work reference with the contractor bid proposal;<br>• The definition of variation is not clear;<br>• Design changes were made by both parties concurrently |
| Terminal Ultimate Soekarno Hatta Airport Project | Angkasa Pura II standard DB Contract | • The unclear definition of variation and design process procedure (changing between basic and detail design)<br>• Design changes were made frequently based on terms of work reference and criteria performance, causing significant volume difference between document contract and installed material. |
| MRT Project | FIDIC Yellow Book 1999 standard DB Contract | • Dispute related to the late issuance of payment certificates<br>• Dispute related of the administration and VO payment |
| Project Title | Type of Contract | Problems |
| Non Tol Over-pass Road Project | Indonesian Ministry of Public Works and Housing standard DB Contract | Different measurement volume between contract document and installed work |
| Asian Games XVIII 2018 Project | Indonesian Ministry of Public Works and Housing standard DB Contract | The reasonableness of the project price value is Rp 5.3 trillion, which is caused by the initial work of the PPK Ministry of PUPR not implementing the RAB used in determining the Self Estimated Price (HPS). |

**Table 4.** Problems arising from the implementation of design and build based on literature review.

| Problem | Problem Identification | Reference |
|---|---|---|
| Construction/Technical | • Difficulties/delay in availability of materials, equipment and labor<br>• Coordination issues arising from design and construction concurrence<br>• Contract change orders | Ibrahim Adamu, Mohammed Awal Sidik, Osei-Tutu Ernest (2017) [17]; O. E. Ogunsanmi, O. A. Salako, and O. M. Ajayi (2011) [18]; Allen Faught and Daniel Tran (2015) [19] |
| Design | • Design changes<br>• Design errors or commissions<br>• Detailed project information | O. E. Ogunsanmi, O. A. Salako, and O. M. Ajayi (2011) [18]; Allen Faught and Daniel Tran (2015) [19]; Ilori, Babsola Olubunmi and Talukhaba, Alfred Atsango (2017) [20] |
| Project Risk | • Different the attitude of decision makers towards risk<br>• Financial risk based on selected contract types | O. E. Ogunsanmi, O. A. Salako, and O. M. Ajayi (2011) [18]; Titouan Plusquelleq, Nadia Lehoux, and Yan Cimon (2016) [21]; Tsung-Chieh Tsai & Min-Lan Yang (2010) [22] |
| Management/Organization (Administrative, Information, Risk) | • Communication system<br>• Control over supply management<br>• Contracting management<br>• Financial failure | O. E. Ogunsanmi, O. A. Salako, and O. M. Ajayi (2011) [18]; Ilori, Babsola Olubunmi and Talukhaba, Alfred Atsango (2017) [20]; Ibrahim S. Abotaleb, A.M. ASCE; Islam H. El-adaway, F. ASCE; and Mohamed B. Moussa (2019) [23] |
| Human Resources (Motivation, Knowledge, Skill, Culture) | • Different perspective view from owner and services provider<br>• Contractor's & client's competencies<br>• Gap in the body of knowledge as related to stakeholders<br>• A lack of understanding of the design-build | Jiyong Ding , Na Wang, and Leichuang Hu (2018) [24]; Titouan Plusquelleq, Nadia Lehoux, and Yan Cimon (2016) [21]; Ibrahim S. Abotaleb, A.M.ASCE; Islam H. El-adaway, F.ASCE; and Mohamed B. Moussa (2019) [23]; Ilori, Babsola Olubunmi and Talukhaba, Alfred Atsango (2017) [20]; Allen Faught and Daniel Tran (2015) [19] |

Based on the results of the explanation above related to the problems of implementing the design and build contract in Indonesia, then for further research the proposed conceptual framework of the development of the SMM for design and build contracts in subsequent studies begins with determining the type of contract and the type of project used as the study boundary. For a clearer picture can be seen in Figure 2 below.

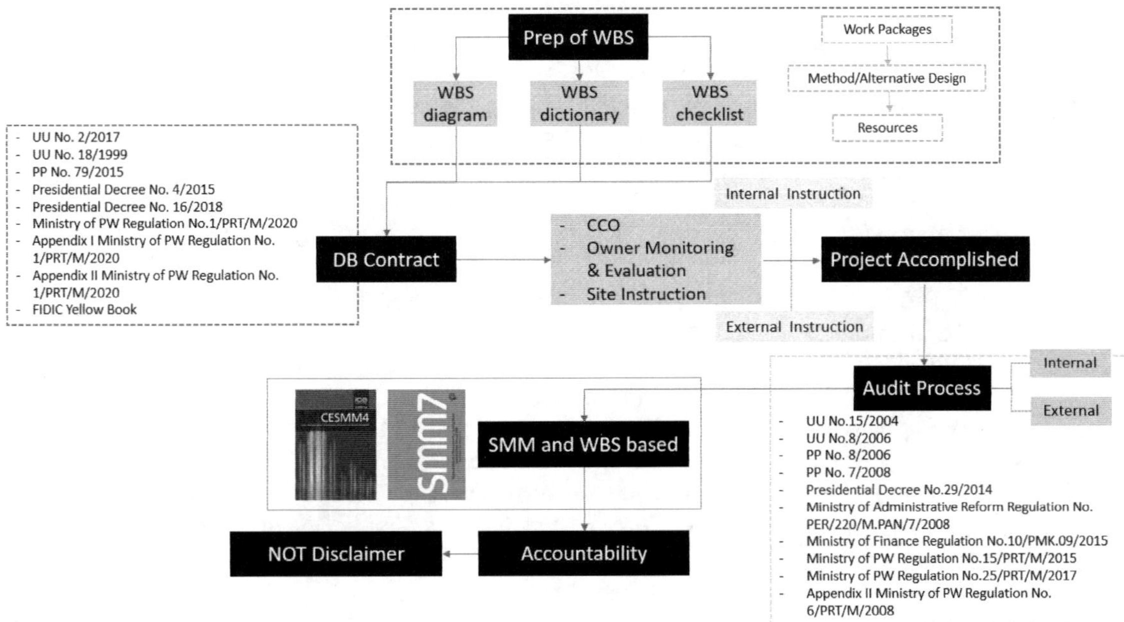


**Figure 2.** The conceptual framework for developing SMM in Indonesia.

## 4. Conclusion

Design and build contract is a procurement method that provides a single responsibility point for a contractor. In this case, the contractor is required to be responsible for meeting all the needs of the owner in terms of design, material requirements and the method of carrying out the work. In the design and build contract the owner can contract directly with the contractor without going through an intermediary party.

The advantage of implementing a DB contract is to utilize the lump sum payment method to transfer all risks to the service provider and streamline the process of procuring goods/services. Moreover, there is flexibility in adjusting the volume of work, because what is stipulated in the contract document is only the design criteria (basic design or concept design). However, in addition to the benefits provided, there are also losses arising from the implementation of a DB contract, one of which is a change in the project due to problems in the specifications, and a long revision process.

The implementation of design and build method impacts Indonesia by providing time cuts in the initial phase of project formation and by making projects faster and completion on time. But in its actuality, there are still many delays in several projects despite implementing the design and build contract. It also turns out that by applying design and build contracts in several state-owned projects, many problems arise, especially with the state institutional auditor.

Several points of shortcoming that have been mentioned become a problem in the implementation of design and build contracts. This creates obstacles in the continuity of project implementation, which is feared to not allow achievement of project objectives. These objectives are defined that the project is completed on time, does not exceed the specified budget and meets the quality requirements and later this research can help further research than this research on the development of WBS-based SMM in Indonesia.

**Acknowledgments**
The authors would love to give thanks to the members of the SMM Team (Amelia Anggraini and Anna Wahidati) who have helped researcher to provide support within the kind of ethical support in finishing this research.

IOP Conf. Series: Materials Science and Engineering **897** (2020) 012016 doi:10.1088/1757-899X/897/1/012016

**References**

[1] Saadin S, Endut I R, Samah S A A and Rizduan A R M 2016 *Int. Academic Research J. of Bus. and Technol.* **2** 51-56
[2] Azizan M A and Ibrahim F A 2015 *Adv. Environ. Biol.* **9** 108-111
[3] Gambo M M and Gomez C P 2016 *J. Eng. Technol.* **6** 144-154
[4] Hale D R, Shrestha P P, Gibson G E and Migliaccio G C 2009 *J. Const. Eng. Manage.* **135** 579-587
[5] Varma V 2006 *2006 Annual Conf. and Exposition* vol 1 (Illinois: American Society for Engineering Education) pp 1-8
[6] Ling F Y Y and Leong E F K 2014 *The Australian J. of Const. Economics and Building* **2** 37-46
[7] Ndekugri I and Turner A 1994 *J. Const. Eng. Manage.* **120** 243-256
[8] Bennet J, Pothecary E and Robinson G 1996 *Designing and Building a World-Class Industry* (Reading: Centre for Strategic Studies in Construction)
[9] Konchar M and Sanvindo V 1998 *J. Const. Eng. Manage.* **124** 435-444
[10] Albuquerque A E C, Primo M A and Pereira F A 2015 *Rev. Bus. Manage.* **17** 822-838
[11] Park M, Ji S H, Lee H S and Kim W 2009 *J. Const. Eng. Manage.* **135** 1125-1137
[12] Seng N W and Yusof A M 2006 *Proc. of the 6th Asia-Pacific Structural Engineering and Construction Conf. (APSEC 2006) 5-6 September 2006 Kuala Lumpur Malaysia* (Kuala Lumpur: APSEC) pp C1-C11
[13] Ashworth A 2006 *Contractual Procedures in The Construction Industry* (United State: Pearson/Prentice Hall)
[14] Morledge R, Smith A and Kashiwagi D T 2008 *Building Procurement* (Oxford: Blackwell Science)
[15] Knight A, Griffith A and King A 2002 *Manage. Decision* **40** 655-662
[16] Chang A S, Shen F Y and Ibbs C W 2010 *Canadian J. of Civil Eng.* **37** 1525-1534
[17] Ibrahim Adamu, Mohammed Awal Sidik, and Osei-Tutu Ernest 2017 *Int. J. Const. Eng. Manage.* **6** 148-159
[18] O. E. Ogunsanmi, O. A. Salako, and O. M. Ajayi 2011 *J. Eng. Project. Prod. Manage.* **1** 46-60
[19] Allen Faught and Daniel Tran 2015 *Proc. of the 5th International Construction Specialty Conference of the Canadian Society for Civil Engineering (ICSC) 8-10 June 2015 Vancouver B.C.* (Vancouver: ICSC) pp 241-1 – 241-9
[20] Ilori, Babsola Olubunmi and Talukhaba, Alfred Atsango 2017 *Int. J. Const. Supply Chain Manage.* **7** 110-123
[21] Titouan Plusquellec, Nadia Lehoux, and Yan Cimon 2016 *Design-build in Construction: Performance and Impact on Stakeholders* (Canada: CIRRELT)
[22] Tsung-Chieh Tsai and Min-Lan Yang 2010 *J. Operation Research Society of Japan* **53** 20-39
[23] Ibrahim S. Abotaleb, A.M.ASCE; Islam H. El-adaway, F.ASCE; and Mohamed B. Moussa 2019 *J. Manage. Eng.* **35** 1-17
[24] Jiyong Ding, Na Wang, and Leichuang Hu 2018 *Advances in Civil Engineering* **2018** 1-14

# Decision-making in the commercial space service design of transportation architecture

**Chia-Jung Shen*** **and Yen Hsu**

The Graduate Institute of Design Science, Tatung University. No.40, Sec.3, Zhongshan N.Rd., Taipei, 10452, Taiwan

*Email: t929040@gmail.com

**Abstract.** Transportation architecture is common globally. The design of rest areas, which allow travelers to rest and refuel, is crucial in such architecture. The design focus of rest areas in Taiwan has shifted from function-oriented features to that meet travelers' diverse needs. Space planning in such areas can be used to integrate highway information, dining services, and leisure activities for travelers; design spaces with local characteristics; and improve service design and innovative service models. This study analyzed the effect of evaluation criteria weights used in highway rest area design tendering in Taiwan on the selection of space designers. The modified Delphi method was used to conduct interviews with experts and select criteria for commercial service space design from a consensus. Subsequently, an analytic hierarchy process was adopted to determine the relative weight of each criterion, and a consistency analysis was conducted. These design criteria were deemed essential by the experts and are thus beneficial for future designers in understanding the aspects in the spatial planning for highway rest areas and acquiring transportation architecture tenders. The results of this study can facilitate identification of highway rest area design criteria, which can serve as a reference for space designers to make related decisions.

Keywords: transportation architecture, commercial space, modified Delphi method, hierarchical analysis, design decision-making

## 1. Introduction

Highways are a primary means of transportation in Taiwan and feature service areas for travelers to rest and refuel. In addition to providing "affective services," highway rest areas have experienced diverse development, with some even transforming into tourist attractions [1]. This study used "transportation architecture" to describe highway rest areas because it more closely reflects the meaning of such areas and encompasses their hardware and software components. In addition to introducing international brands to shops, highway rest areas incorporate services such as providing local tourism information and selling local souvenirs. The interior and exterior design of highway rest areas also integrates local cultural elements to drive local industry development. Regarding shop management in highway rest areas, training of retail staff members and food safety management have become stricter.

In Taiwan, old buildings at highway rest areas are met with challenges concerning maintenance and upgrade design. When designing a highway rest area, shop managers and space designers must account for travelers' needs, the amount of investment funds, and related costs. Identifying key design factors is time and cost intensive. Therefore, design decision makers are particularly crucial to the renovation of highway rest areas. Such designers must work under the premise of old building maintenance and innovative design and also account for the design concepts of complex expert systems and circular business models for sustainable development.

The planning of commercial spaces in transportation architecture differs from that of general shopping malls in terms of user purposes and length of stay. Such commercial spaces excel in disseminating local tourism information compared with other media. Therefore, space planning must entail specific objectives and criteria and also incorporate local cultural elements to showcase local characteristics (Figure 1). This promotes local marketing and tourism activities and is considerably affected by innovative space design. Figure 1a presents the interior space of the mall at the Southbound Rende Service Area. The theme of this rest area is "tradition and innovation;" hence, retro elements such as wooden farming boxes are used to decorate the ceiling, and bamboo furniture is used in the dining area. Figure 1b presents the northbound rest area with the theme of "technology takeoff." Hence, interactive projection technology is used to introduce local attractions, and the ceiling features astronomical elements of the Tainan Astronomical Education Area. The rapid development of the Taiwanese design industry has prompted various industries to invest in design-related operations. Design criteria in planning commercial space enable professional designers to improve their evaluation scores when competing for tender opportunities, thus helping them acquire tenders.

The research motivation of this study originated from the authors' experiences in encountering various difficulties when devising design plans. This study focused on the design criteria for planning commercial space in transportation architecture (i.e., highway rest areas) to achieve the following objectives: (1) identify design criteria while considering various factors related to design, management, and profit, (2) combine the modified Delphi method (MDM) and hierarchical analysis to explore the adequacy of the devised comprehensive research framework, and (3) determine the weighted ranking of design criteria according to the opinions of relevant experts, space designers, and managers. The study results can serve as a reference for the design and planning of commercial service space.

(a) (b)

**Figure 1.** (a) Combining space design with theme display (simulation image of space inside the Southbound Rende Service Area). (b) Combining space design with innovation design and local revitalization (simulation image of space inside the Northbound Rende Service Area).

## 2. Literature review

Transportation architecture refers to any structure that facilitates passenger or freight transportation and encompasses various facilities, including airports, harbors, roads, and railroads. In Taiwan, highway rest areas are comprehensive civil engineering buildings that not only have diverse transportation and service functions but also ensure smooth traffic and road safety. Since 1980, the Freeway Bureau of Taiwan has commissioned private companies to manage highway rest areas. The selection of such companies is based on the assessment method of the most advantageous tender. Currently, the selection focus has shifted from profit oriented to service oriented. Moreover, the Bureau has conducted user satisfaction surveys for several years to implement improvement measures. Companies selected to plan highway rest areas are now required to design affective services instead of basic services [1]. In recent years, the environmental planning of highway service areas has integrated aesthetic concepts and created a new trend of transforming such areas into tourist attractions [2]. Because of its popularity among younger citizens, social media check-ins has become an effective marketing strategy. In response to overall user needs, highway rest areas should enhance their recreational functions and integrate multifunctional malls, food courts, multimedia services, book-sharing stands, Sunday markets, and various live performances that promote local culture.

Design- or innovation-oriented corporations or services can establish environments that are conducive to creativity and innovation. These environments provide such corporations advantages over their competitors and enhance corporate performance and strategies [3]. Innovation in space design encourages consumption behavior and helps managers achieve their profit goals [4].

Space design in malls entails various methods of incorporating local characteristics and creating empathetic links with local people. Through space design, local features and social consciousness are realized. Therefore, mall spaces must be adjustable and modularized to enable flexible and dynamic spatial adjustment [5].

In the analysis of design elements in mall space planning, mall designers and managers must agree on the rationality of line planning. Facility convenience and refined space allocation affect consumers' willingness to visit a mall [6]. The aforementioned literature review indicated that the various dimensions of space planning in malls must be weighted. Organizing or executing a design project requires the interdisciplinary integration of expert systems in various fields. These expert systems are assigned to various stages of the development process for multidimensional assessment. However, due to budget and scheduling limitations, crucial design elements must be prioritized in the planning and construction of transportation architecture. Experts from various professional fields, academics, and related brand managers agree that space design entails various essential criteria dimensions, namely hardware design, local feature design, service design, and management strategies. Key factors must be identified for each dimension in a hierarchical framework and serve as an effective reference for devising the overall design plan.

## 3. Research methods

The research procedures of this study were as follows: (1) Data related to the development of and construction white papers on Taiwanese highway rest areas were collected to compile relevant theories and discourses, thereby establishing the overall research framework. (2) Key design criteria were proposed on the basis of literature, theories, design elements, and strategies for planning commercial space. (3) According to the competitive advantages of commercial space design and literature related to highway rest areas, the MDM was used to invite experts, who shared their experience, knowledge, and opinions, to reach a consensus and identify competitiveness criteria for planning commercial spaces in transportation architecture. The validity and reliability of each criterion were verified, and factor analysis was performed. (4) An analytical hierarchy process (AHP) questionnaire was distributed for professional assessment, to evaluate consistency in the expert opinions, and determine relative weights of essential design criteria. The results can serve as a reference for space designers to acquire tenders for designing highway rest areas.

### *3.1. MDM application*

The MDM originated from the conventional Delphi method, also known as the expert judgement method. It is a group decision-making approach mostly used in qualitative research. In the MDM, the knowledge and experience of experts are compiled by using inquiries and feedback to minimize variance in opinions. The MDM retains the concept and advantages of the conventional Delphi method. Experts remain anonymous while exchanging opinions and expressing their thoughts. In contrast to the conventional Delphi method, the MDM omits the first-round open-ended questionnaire and simplifies the procedure to a literature review. Hence, researchers need not interpret the open-ended answers provided by experts and can instead focus on researching the topic of interest. This helps researchers reach a consensus among the experts to expedite subsequent operations. The AHP was employed in the second phase of this study. This method is mainly used to examine problems concerning multiple assessment criteria under uncertain or unknown situations. Such situations include application planning, sequential decision-making, alternative plans, and performance evaluation. The AHP is prevalent in marketing, management, and various scientific fields. Figure 2 displays the research procedures of this study.

**Figure 2.** Research procedures

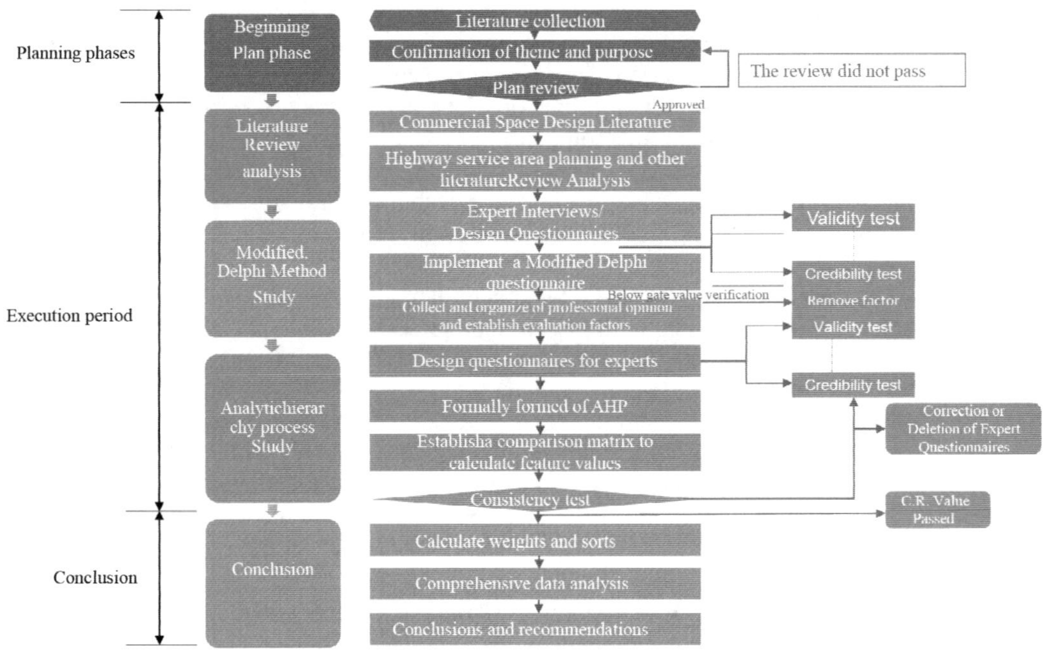


In the planning phase, literature was collected to determine the research direction. Literature review and expert interviews were then conducted during the execution period to compile expert opinions through the MDM. Table 1 lists the design principles and purposes of the questionnaires used in the MDM and AHP. Finally, a decision-making model for planning commercial space service design transportation architecture was created.

**Table 1.** Design principles and purposes of the questionnaires.

| | Design principles | Design purpose | Recovery |
|---|---|---|---|
| First Questionnaire (Expert Interview and Researcher Construction) | According to the relevant highway service investment policy white paper and related commercial space design literature, form the design pointer Seek expert advice Understand the current standard of tenders and tenders in service areas and the requirements for related service proposals | Get expert selection design planning unit design pointer view | Send out 20 questionnaires, 20 points, recovery rate of 100%. |
| Second Questionnaire | Using hierarchical analysis, a presentation of commercial space design guidelines for each service area | Get the relative weight value of each pointer | Send 30 copies, recycle 30 copies, 100% recovery rate |

*3.2. Questionnaire design*

The design principles of the questionnaires were as follows: (1) Literature related to design evaluation strategies, space-design methods, and building construction regulations was reviewed to create a preliminary list of space-design criteria. (2) Various experts were invited to select suitable criteria from

IOP Conf. Series: Materials Science and Engineering **897** (2020) 012017 doi:10.1088/1757-899X/897/1/012017

the questionnaires. These experts included representatives from interior design and construction associations, architects and technicians from architecture associations, and scholars from design departments. (3) The selected criteria were categorized to compile a list suitable for planning commercial spaces in highway rest areas. Accordingly, the formal AHP questionnaire was formulated. Based on the questionnaire responses, a weight matrix was created to perform comprehensive data analysis and propose conclusions and suggestions.

### *3.3. AHP*

The AHP was developed in 1971 by Saaty, a scholar at the University of Pittsburgh, and is mainly used to develop decision-making evaluation criteria for multiple objectives under uncertain situations. In multiobjective decision-making, decision-makers attempt to manage a complex system that features numerous variables as well as complex and uncertain factors. The goal of the AHP is to categorize complex and unstructured problems into hierarchical levels or structures, and the opinions of experts and decision-makers are compiled to simplify the complex system. Matrices are then calculated to compare criteria in each hierarchical level, and corresponding vectors are determined to rank the criteria, which serve as references for evaluation or decision-making.

Among various systems engineering theories, the AHP is relatively useful for weight calculation. The AHP arranges various factors of a complex problem into hierarchical order and is considered a logical multiobjective, multicriterion decision-making approach that combines qualitative and quantitative analyses.

## 4. Case research and analysis

### *4.1. Design criteria formulation and selection*

Based on literature related to factors affecting the planning of commercial space design [6, 7] and the opinions provided by design experts, construction project-execution units, mall managers, and tender bidders, this study compiled and analyzed relevant data to propose four major design elements and 20 design criteria (Figure 3). Subsequently, these criteria were evaluated using opinion exchange among experts.

**Figure 3.** AHP model for the decision-making criteria of commercial space design in transportation architecture.

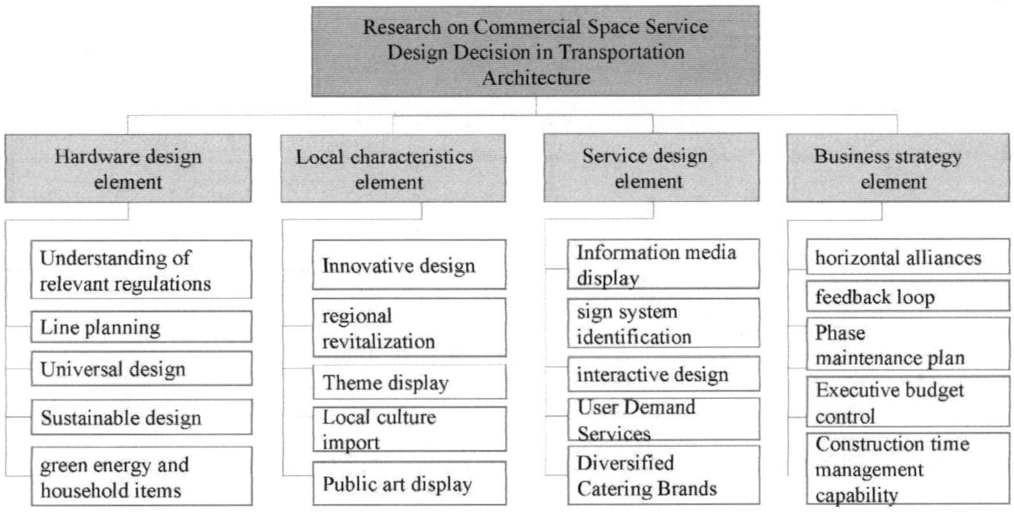


### *4.2. Validity and reliability analysis*

Table 2 lists the results of the Bartlett test of sphericity. The Kaiser–Meyer–Olkin (KMO) value was .828, which was >.8, thus indicating that the collected data were suitable for factor analysis. The chi-square value determined using the Bartlett test of sphericity was 1154.101 and statistically significant (p = 0.000). The questionnaires used in this study were formulated by referencing relevant literature and collecting data from expert interviews; hence, their content validity required verification. Although content validity verification was relatively subjective, the pretest indicated that the questionnaire content was highly valid [8].

**Table 2.** KMO & Bartlett test.

| | | |
|---|---|---|
| Kaiser-Meyer-Olkin Measure of Sampling Adequacy | | .828 |
| Bartlett's Test of Spherical | Approximate Chi-Square | 1154.101 |
| | df | 190 |
| | Sig. | .000 |

Based on the data compiled from the literature review and interviews with experts and scholars in the first study phase, a list of evaluation criteria for planning commercial space in transportation architecture was created (table 3). These criteria were used to develop an MDM expert questionnaire, which contained structured items and involved four elements (i.e., hardware design, local characteristics, service design, and business strategy) and 20 criteria. The questionnaire was used to enable experts and scholars to reach a consensus and evaluate the adequacy of the criteria.

Babakus and Mangold [9] argued that the 5-point Likert scale can enhance the comprehensibility of questionnaire items and the quality of questionnaire responses. Therefore, this scale was employed to evaluate questionnaire items. In the first study phase, 20 questionnaires were distributed and recovered, yielding a recovery rate of 100%. The mean scores of all criteria were ≥3.5, thus indicating that all criteria were considered essential to space design. Furthermore, the standard deviation values of the mean scores were all <1, implying high consistency among experts' opinions.

**Table 3.** First questionnaire results.

| | | Evaluation criteria | M | SD | MO | ∣ MO-M ∣ |
|---|---|---|---|---|---|---|
| 1 | Hardware design element | Understanding of relevant regulations | 4.90 | 0.305 | 5 | 0.100 |
| 2 | | Line planning | 4.97 | 0.183 | 5 | 0.033 |
| 3 | | Universal design | 4.87 | 0.346 | 5 | 0.133 |
| 4 | | Sustainable design | 4.83 | 0.379 | 5 | 0.167 |
| 5 | | green energy and household items | 4.80 | 0.407 | 5 | 0.200 |
| 6 | Local characteristics element | Innovative design | 4.73 | 0.45 | 5 | 0.267 |
| 7 | | regional revitalization | 4.77 | 0.43 | 5 | 0.233 |
| 8 | | Theme display | 4.77 | 0.43 | 5 | 0.233 |
| 9 | | Local culture import | 4.67 | 0.479 | 5 | 0.333 |
| 10 | | Public art display | 4.53 | 0.629 | 5 | 0.467 |
| 11 | Service design element | Information media display | 4.80 | 0.407 | 5 | 0.200 |
| 12 | | sign system identification | 4.93 | 0.254 | 5 | 0.067 |
| 13 | | interactive design | 4.57 | 0.504 | 5 | 0.433 |
| 14 | | User Demand Services | 4.77 | 0.43 | 5 | 0.233 |

| | | | | | | |
|---|---|---|---|---|---|---|
| 15 | | Diversified Catering Brands | 4.53 | 0.507 | 5 | 0.467 |
| 16 | Business strategy element | horizontal alliances | 4.50 | 0.509 | 5 | 0.500 |
| 17 | | feedback loop | 4.37 | 0.669 | 5 | 0.633 |
| 18 | | Phase maintenance plan | 4.67 | 0.479 | 5 | 0.333 |
| 19 | | Executive budget control | 4.77 | 0.43 | 5 | 0.233 |
| 20 | | Construction time management capability | 4.83 | 0.379 | 5 | 0.167 |

### *4.3. AHP questionnaire survey and analysis*

In the second study phase, survey respondents were mostly experts in design-related fields (including architects, interior designers, landscape designers, engineering technicians, brand managers, and academics). A total of 30 questionnaires were distributed and recovered. The questionnaires were primarily distributed through the Internet. Questionnaire responses were subject to AHP analysis to determine the relative weight of each criterion.

**Table 4.** Analytical hierarchy process results.

| Level 1 | Statistic weight | Level 2 Design index factor | Intra-group weights | Relative weight | Sorting weights |
|---|---|---|---|---|---|
| Hardware design element | 0.2588 | Understanding of relevant regulations | 0.0533 | 0.2057 | 7 |
| | | Line planning | 0.0927 | 0.3580 | 1 |
| | | Sustainable design | 0.0427 | 0.1650 | 14 |
| | | Sustainable design | 0.0399 | 0.1543 | 17 |
| | | green energy and household items | 0.0303 | 0.1170 | 20 |
| Local characteristics element | 0.2687 | Innovative design | 0.0726 | 0.2703 | 3 |
| | | regional revitalization | 0.054 | 0.2010 | 10 |
| | | Theme display | 0.0552 | 0.2056 | 8 |
| | | Local culture import | 0.0538 | 0.2004 | 11 |
| | | Public art display | 0.033 | 0.1227 | 19 |
| Service design element | 0.2611 | Information media display | 0.0416 | 0.1593 | 16 |
| | | sign system identification | 0.071 | 0.2719 | 2 |
| | | interactive design | 0.0441 | 0.1691 | 12 |
| | | User Demand Services | 0.0651 | 0.2492 | 4 |
| | | Diversified Catering Brands | 0.0393 | 0.1504 | 18 |
| Business strategy element | 0.2114 | horizontal alliances | 0.0351 | 0.1660 | 13 |
| | | feedback loop | 0.0347 | 0.1642 | 15 |

| | | | | |
|---|---|---|---|---|
| | Phase maintenance plan | 0.0492 | 0.2326 | 6 |
| | Executive budget control | 0.0494 | 0.2335 | 5 |
| | Construction time management capability | 0.0431 | 0.2038 | 9 |

The weights of hardware design, local characteristics, service design, and business strategy elements were 0.2588, 0.2687, 0.2611, and 0.2114, respectively. This indicates that when designing commercial spaces of transportation architecture, the local characteristics element is the most essential, followed by service design, hardware design, and business strategy. Within the hardware design element, understanding of relevant regulations and line planning yielded a weight of >0.05. Within the local characteristics element, experts emphasized innovative design, regional revitalization, theme display, and local culture import. Within the service design element, sign system identification and user demand services were considered the most crucial criteria. In the business strategy element, no criteria were considered essential.

Of all the criteria, the top five were line planning, sign system identification, innovative design, user demand services, and executive budget control (Figure 4). Of all the elements, criteria with weights of >0.05 were understanding of relevant regulations, line planning, innovative design, regional revitalization, theme display, local culture import, sign system identification, and user demand services. The experts regarded these criteria as essential (Figure 4).

**Figure 4.** Results of group decision-making (weights) by experts.

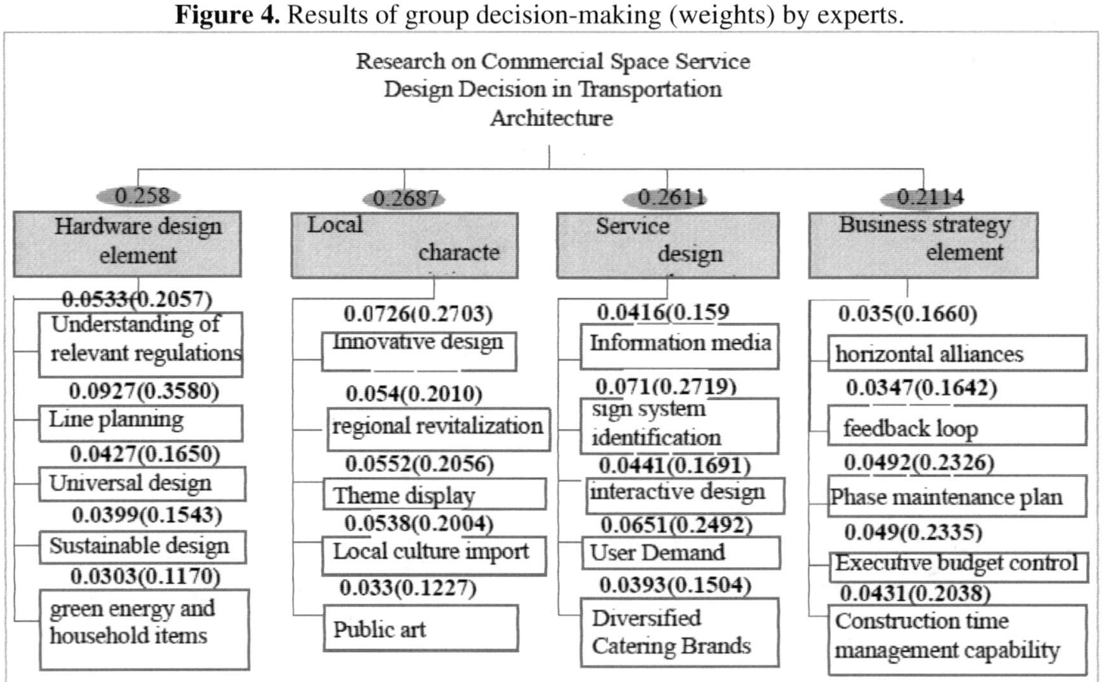


## 5. Conclusions and limitations

Decision-making problems are frequently encountered in real-world situations. However, using quantitative methods to describe related factors in this type of problem is often difficult. The AHP is effective for solving such problems. This process organizes a complex decision-making system into hierarchical levels. Related factors in each level are then compared in terms of their importance to provide a quantitative reference for making decisions. The space design industry in Taiwan has become increasingly diverse. In response to consumers' increasing standards for design quality and the perpetually competitive market, designers must have professional abilities across different fields to satisfy consumer needs. These abilities include understanding related regulations, material selection, and cultural innovation. This study compiled a list of criteria for designing commercial spaces in transportation architecture and ranked the criteria by their weights. The study results can serve as a reference for future designers to make adequate decisions in designing commercial spaces in highway rest areas, thereby helping them acquire tenders for related projects.

The research limitations of this study were as follows:

(1) The questionnaires used in this study were based on the MDM. Future researchers should use scales based on service design and innovation theories to acquire more objective data and create innovation-design models.

(2) The questionnaires used in this study were designed on the basis of the space-design experience of experts and related literature and theories. To ensure the integrity of questionnaire items and respondents' willingness to complete the questionnaires, the wording used in the items must be clear, and the number of items must be reasonable. Consequently, items related to some topics were omitted from the questionnaires.

(3) This study mainly focused on commercial space in transportation architecture. Future scholars may extend the results of this study by targeting commercial spaces in other types of transportation architecture, such as airports or railway stations.

## References

[1] Hsu CH, Lin KH 2005 Exploration on the development of service sections on the highway-taking no. 2 highway as an example *J Sport Recreat Manag* **2** 118–132.

[2] Liu HC, Lei WY, Shang YW 2012 A study on passengers' cognition of needs and perceived experiences on service area of the national freeway *NCYU Phys Educ Health Recreat J* **11** 13–25.

[3] Tidd J, Bessant J 2009 *Managing Innovation: Integrating Technological, Market and Organizational Change* (Hoboken: Wiley).

[4] Johne A 1999 Using market vision to steer innovation *Technovation* **19** 203–7.

[5] Nystrom M, Folke C 2001 Spatial resilience of coral reefs *Ecosystems* **4** 406–417.

[6] Lin MF, Perng YH 2019 Exploration of key commercial space design indicators *Int Conf on Civil Engineering and Materials Science* (Bangkok, Thailand) pp. 324–331.

[7] Lee CJ, Sung TJ, Cai DC 2012 The servicescape attributes and consumer perceived qualityevidence from theme restaurants *J Des* **17** 1–20.

[8] Yang MW, Ho MC 2005 An empirical study on factors affecting job market of visual communication design professionals *J Des* **10** 1–19.

[9] Babakus E, Mangold WG 1992 Adapting the SERVQUAL scale to hospital service: an empirical investigation *Health Serv Res* **26** 767–786.

# AUTHOR INDEX

**Institute of Physics**
Dirac House, Temple Back
Bristol BS1 6BE UK

ISSN: 1757-8981
ISBN 978-1-7138-2811-2

9 781713 828112

CURRAN ASSOCIATES INC.
proceedings.com